Lecture Notes in Mathematics

Edited by A. Dold, Heidelberg and B. Eckmann, Zürich

Series: Australian National University, Canberra
Advisers: L. G. Kovács, B. H. Neumann, and M. F. Newman

374

Sadayuki Yamamuro

Differential Calculus
in Topological Linear Spaces

Springer-Verlag
Berlin · Heidelberg · New York 1974

AMS Subject Classifications (1970): 46-02, 46B99, 46G05, 58-02, 58C20

ISBN 3-540-06709-4 Springer-Verlag Berlin · Heidelberg · New York
ISBN 0-387-06709-4 Springer-Verlag New York · Heidelberg · Berlin

This work is subject to copyright. All rights are reserved, whether the whole or part of the material is concerned, specifically those of translation, reprinting, re-use of illustrations, broadcasting, reproduction by photocopying machine or similar means, and storage in data banks.
Under § 54 of the German Copyright Law where copies are made for other than private use, a fee is payable to the publisher, the amount of the fee to be determined by agreement with the publisher.
© by Springer-Verlag Berlin · Heidelberg 1974. Library of Congress Catalog Card Number 73-21376. Printed in Germany.
Offsetdruck: Julius Beltz, Hemsbach/Bergstr.

CONTENTS

Introduction		1
Chapter 1.	Definitions and fundamental properties	6
§1.1	Directional derivatives	6
§1.2	M-derivatives	7
§1.3	Mean value theorems	13
§1.4	Relations among M-differentiabilities	17
§1.5	Differentiability in spaces with projective topology	19
§1.6	Differentiability in inductive limits of increasing families of subspaces.	21
§1.7	Differentiability and continuity	23
§1.8	Higher derivatives	25
§1.9	Equicontinuously differentiable mappings	30
§1.10	Uniform differentiability	34
§1.11	Partial derivatives	36
§1.12	Other differentiabilities	39
Chapter 2.	Compact mappings	43
§2.1	Compact mappings and Fréchet derivatives	43
§2.2	Compact mappings and Hadamard differentiability	47
Chapter 3.	Inverse mapping theorems	50
§3.1	Differentiation in $L(E,F)$	50
§3.2	Differentiability of inverse mappings	53
§3.3	The space $L_p(E,F)$	58
§3.4	C_p-mappings and an inverse mapping theorem	61
§3.5	Other theorems on inverse mappings	69

Chapter 4.	Differentiability of semi-norms	76
§4.1	Hadamard differentiability of semi-norms	76
§4.2	Fréchet differentiability of semi-norms	81
§4.3	Higher derivatives of semi-norms	84
§4.4	Differentiability of the supremum norms of function spaces	87
§4.5	Differentiability of norms of L_p-spaces	95
Chapter 5.	Smoothness	98
§5.1	S-categories	98
§5.2	S-smooth spaces	100
§5.3	Partitions of unity	106
Chapter 6.	Differentiability of mappings of a real variable	111
§6.1	Differentiability of Lipschitz mappings	111
§6.2	Differentiability of Stepanoff mappings	114
§6.3	Theorems of L. Schwartz and A. Grothendieck	121
Chapter 7.	Sets of differentiable mappings	124
§7.1	Idempotents of semigroups of differentiable mappings	124
§7.2	Automorphisms of semigroups of differentiable mappings	126
§7.3	Near-rings of differentiable mappings	132
Appendix 1.	Sequential spaces	140
Appendix 2.	Continuity of composition mappings	143
Appendix 3.	Differentiability of inverse mappings	147
List of symbols		155
References		159
Index		177

INTRODUCTION

Throughout this note, all linear spaces are those which are over the real number field R. We shall use Greek letters to denote real numbers.

For topological linear spaces E and F, let L(E,F) be the set of all continuous linear mappings of E into F.

Assume that E and F are normed linear spaces and A is an open subset of E. Then, a mapping $f : A \to F$ is said to be <u>Fréchet differentiable at</u> $a \in A$ if there exists $u \in L(E,F)$ such that the "<u>remainder of</u> f <u>at</u> a" :

$$r(f,a,x) = f(a+x) - f(a) - u(x)$$

satisfies the following condition :

$$\lim_{\|x\| \to 0} \|x\|^{-1} r(f,a,x) = 0.$$

The linear mapping u is uniquely determined and is denoted by $f'(a)$, which is called the <u>Fréchet derivative of</u> f <u>at</u> a. This notion of differentiability is the basis of the differential calculus in normed linear spaces, and any good definition of differentiability in topological linear spaces must coincide with it when the spaces are normed.

According to V.I. Averbukh and O.G. Smolyanov [2], more than twenty such definitions had appeared by the time when their paper was written. In the paper, which contains skillfully constructed examples, they have tidied up this chaos and classified these into less than ten groups (see §1.12). We all may agree, however, that there are still too many differentiations.

In this note, we shall focus our attention to two definitions: the weakest and the strongest among those definitions which coincide with the Fréchet differentiability in the case when the spaces are normed. They will be called the <u>Fréchet</u> and the <u>strong</u> <u>Fréchet differentiability</u> respectively.

The Fréchet differentiability of $f : A \to F$ is defined in terms of uniform convergence of $\{\varepsilon^{-1} r(f,a,\varepsilon x)\}$ on bounded sets with respect to x when $\varepsilon \to 0$

(see §1.2). If we recall the importance of this convergence in the theory of topological linear spaces, we realize that this is the most natural generalization of Fréchet differentiability of normed spaces.

The strong Fréchet differentiability of $f : A \to F$ is defined in terms of uniform convergence of $\{\varepsilon^{-1}r(f,a,\varepsilon x)\}$ on a neighbourhood of zero with respect to x (see §1.7). This convergence also appears in the theory of topological linear spaces; for example, in the definition of completely continuous mappings.

If E and F are normed linear spaces, the convergence

$$\lim_{\|x\| \to 0} \|x\|^{-1} r(f,a,x) = 0$$

is equivalent to

$$\lim_{\varepsilon \to 0} \sup_{\|x\| \leq 1} \|\varepsilon^{-1} r(f,a,\varepsilon x)\| = 0$$

(see (1.2.6)), namely, the uniform convergence on the unit ball, which is bounded as well as a neighbourhood of zero. If a topological linear space is not normed, no neighbourhood of zero is bounded. Therefore, in order to have a differential calculus which effectively generalizes the normed space calculus, we need at least two differentiabilities; one that is defined by the uniform convergence on bounded sets and another defined by the uniform convergence on a neighbourhood of zero.

In Chapter 1, fundamental properties of these differentiations will be given. Particularly, we would like to call attention to the following three points:

(1) The Fréchet differentiability at a point does not imply continuity at the point (see §1.7). However, it satisfies the chain rules of all orders.

(2) The strong Fréchet differentiability at a point implies continuity at the point (see §1.7). However, it has serious difficulty in connection with the chain rule (see Note of §1.8 and Appendix 2).

(3) Since we need a differentiability which implies continuity and enjoys chain rules of all orders in Chapter 5, we propose the equicontinuous differentiability.

In Chapter 2, we consider the relation between the compactness of a differentiable mapping and that of its derivative. As to the importance of this problem in the so-called variational methods in nonlinear analysis, see the book of Vainberg [1].

In Chapter 3, we first consider the composition mapping of $L(E,F) \times L(F,G)$ into $L(E,G)$ for topological linear spaces E, F and G. It turns out to be discontinuous but infinitely times Fréchet differentiable. Since it is not continuous, it is not strongly Fréchet differentiable.

Next, we consider the differentiability of the inverses of differentiable mappings. We shall give only elementary properties there and deeper investigations will be done in Appendix 3. We see from these results that Fréchet differentiability is anything but satisfactory when the differentiability of the inverse mapping is involved. The strong Fréchet differentiability behaves better, but even it falls far short of the requirement for the inverse mapping theorem: the most important bridge between topological and differentiable investigations. It has become clear that it is impossible to obtain a faithful generalization of the normed space version by using any of the existing definitions of differentiability. Then what kinds of additional assumptions are needed to make it possible? We present a sample in §3.3 and §3.4.

In Chapter 4, we study the differentiability of semi-norms, which is the first step towards the study of differentiable structure of topological linear spaces. Here, for the first time, the Hadamard differentiability plays a major rôle. If a semi-norm is Gâteaux differentiable at a point, it is Hadamard differentiable there. Moreover, unlike the Gâteaux differentiation, the Hadamard differentiation enjoys chain rules of all orders. In other words, results on Gâteaux differentiability of semi-norms can be brought into categorical argument by replacing it by Hadamard differentiability. This is the most remarkable benefit we get from this differentiation.

Chapter 5 is devoted to construct a topological linear space version of the S-category theory proposed first by R. Bonic and J. Frampton [2]. The results in

Chapter 4 are used to determine various kinds of smoothness of various topological linear spaces. Only (5.3.5) will be enough to illustrate the power of this theory.

We have been working for several years on the generalization of the following result of K.D. Magill, Jr. [1]: every automorphism of the semigroup of all differentiable functions of R into R is inner. As we shall explain in §7.2, the main part of the proof is to prove that one particular mapping is differentiable in the sense by which the given semigroup is defined. The proof, for some special cases where we have been successful, starts by proving that the mapping is differentiable in a very weak sense. In Chapter 6, we have collected various results of such character.

Non-linear version of the theory of Banach algebras has not been started yet. It will eventually need deeper understanding on the interactions between analysis and algebraic theories like semigroups and near-rings. We end this note with a short account on somewhat obvious results in this direction.

As the substitutes of Fréchet differentiability in normed spaces, we have chosen two (Fréchet and strong Fréchet) differentiabilities, and we have put more assumptions on them according to the problem under consideration. This could be one way to overcome special difficulties that arise in non-normed cases.

If we want to construct more monotheistic calculus, we seem to have three ways. The first is to restrict the class of topological linear spaces. We know that, if we consider only sequential spaces, difficulties (1) and (2) mentioned above disappear (see (1.7.1), Note of §1.8 and (3.2) in Appendix 2). This will cover the spaces which appear in the theory of distribution. The second way is to go out of the traditional frame of topological linear spaces. Two such works have appeared in this Lecture Note series. One is A. Frölicher and W. Buchner [1], where a calculus on pseudo-topological linear spaces or a linear space with limit structure has been constructed. The other is U. Seip [1], where a calculus was constructed on a class of topological and linear spaces which is cartesian closed as a category (compactly generated spaces). The third way is to employ new notions only when it becomes necessary. Recently, Professor H.H. Keller has kindly informed us that he had constructed a theory of C^p-functions between locally convex spaces, using limit structures only in the spaces of multilinear mappings; the notion of differentiable functions with domain and image in a limit vector space is not needed. This method appears to be the most natural one by which the future calculus will be constructed.

For the history of differential calculus in topological linear spaces we refer to Averbukh-Smolyanov [2]. Nashed [1] contains an extensive bibliography and survey as well as history.

In the following, [i,j,k] will denote a definition given in Chapter i, Section j and it is the k-th statement. The propositions and theorems will be denoted by (i,j,k).

About three years ago, two new graduates started their postgraduate works in which differentiability in topological linear spaces was an indispensable tool. To help their works as a supervisor, I have set out to collect relating materials. It turned out, however, that it was I who profited most from this collaboration, of which this note is the product. My deepest thanks go to them : Dr. John W. Lloyd and Dr. Graham R. Wood.

I thank Mrs. A. Zalucki for typing this manuscript.

I also thank Dr. R.J. Loy for useful advices, linguistic as well as mathematical, and Dr. M.F. Newman for his kind considerations.

I owe too much to Professor Bernhard H. Neumann to express my thanks in a few lines.

CHAPTER 1. DEFINITIONS AND FUNDAMENTAL PROPERTIES

The aim of this chapter is to give definitions of differentiabilities and their fundamental properties. Although these will be given in the case of topological linear spaces (TLS) over the real number field R, the case of normed linear spaces (NLS) over R is the most important one and will be treated as such.

Throughout, if $E \in$ TLS, we shall denote the fact that a subset A of E is open by $A \in \mathcal{O}(E)$. For the symbols like these, see the list of symbols at the end of this lecture.

§1.1 Directional derivatives

In this section, we assume that

$$E, F \in \text{TLS}, \quad a \in A \in \mathcal{O}(E) \quad \text{and} \quad f, g : A \to F.$$

[1.1.1] f is said to be <u>differentiable at</u> a <u>in the direction of</u> $x \in E$ <u>if</u> the limit

$$f'(a,x) = \lim_{\varepsilon \to 0} \varepsilon^{-1}[f(a+\varepsilon x) - f(a)]$$

exists. If this is the case, we write

$$f \in D(a, F; \to x)$$

and $f'(a,x)$ will be called <u>the directional derivative of</u> f <u>at</u> a <u>in the direction of</u> x. We put

$$D(a, F; \to E) = \bigcap_{x \in E} D(a, F; \to x).$$

The following statements follow immediately from the existence of the limits.

(1.1.2) <u>If</u> $f, g \in D(a, F; \to x)$, <u>then</u>

1°. $f \in D(a, F; \to \xi x)$ <u>for all</u> $\xi \in R$ <u>and</u>

$$f'(a, \xi x) = \xi f'(a, x).$$

2°. $\alpha f + \beta g \in D(a,F;\to x)$ <u>for any</u> $\alpha, \beta \in R$ <u>and</u>

$$(\alpha f+\beta g)'(a,x) = \alpha f'(a,x) + \beta g'(a,x).$$

On the other hand, there is no connection between the continuity and the linearity of $f'(a,x)$ with respect to x.

(1.1.3) <u>Assume that</u> $f \in D(a,F;\to E)$.

1°. <u>The continuity of</u> $f'(a,x)$ <u>with respect to</u> x <u>does not imply its linearity</u>.

2°. <u>The linearity of</u> $f'(a,x)$ <u>with respect to</u> x <u>does not imply its continuity</u>.

PROOF. 1°. Consider $f : R^2 \to R$ defined by

$$f(\xi, \eta) = \eta^3/(\xi^2 + \eta^2).$$

2°. Any linear non-continuous mapping is an example.

(1.1.4) <u>Let</u> $F \in LCS$ <u>and</u> $f \in D(a,F;\to x)$. <u>Then, for any</u> $\bar{a} \in \bar{F}$, <u>the function</u> $\phi : R \to R$ <u>defined by</u>

$$\phi(\xi) = <f(a+\xi x), \bar{a}>$$

<u>is differentiable at</u> $\xi = 0$ <u>and</u>

$$(\tfrac{d}{d\xi}\phi(\xi))_{\xi=0} = <f'(a,x), \bar{a}>.$$

§1.2 <u>M-derivatives</u>

We shall introduce three kinds of differentiabilities : Fréchet, Hadamard and Gâteaux. As before, we assume

$$E, F \in TLS, \quad a \in A \in O(E) \quad \text{and} \quad f : A \to F.$$

[1.2.1] Let M be a class of subsets of E such that every singleton belongs to M. We shall say that f is <u>M-differentiable at</u> a if there exists $u \in L(E,F)$ such that

$$\lim_{\varepsilon \to 0} \varepsilon^{-1} r(f,a,\varepsilon x) = 0$$

uniformly with respect to x on each member of M, where the "remainder" $r(f,a,x)$ is defined by

$$r(f,a,x) = f(a+x) - f(a) - u(x).$$

In this case, we write

$$f \in D_M(a,F)$$

and the continuous linear mapping u is called the M-derivative of f at a.

In other words, $f \in D_M(a,F)$ if and only if f is defined on an open set containing a and, for any $V \in N(F)$ and any $M \in \mathcal{M}$, there exists $\alpha > 0$ such that

$$\varepsilon^{-1} r(f,a,\varepsilon x) \in V \quad \text{if} \quad x \in M \quad \text{and} \quad 0 < |\varepsilon| < \alpha$$

Since $\mathcal{M}$ contains all singletons, we have

(1.2.2) $D_M(a,F) \subset D(a,F;\to E)$.

Therefore, the mapping u is uniquely determined. We denote it by $f'(a)$. The following property is obvious.

(1.2.3) <u>If</u> $f,g \in D_M(a,F)$, <u>then</u> $\alpha f + \beta g \in D_M(a,F)$ <u>for any</u> $\alpha, \beta \in \mathbb{R}$ <u>and</u>

$$(\alpha f + \beta g)'(a) = \alpha f'(a) + \beta g'(a).$$

Now, we define three differentiabilities. For the symbols, see the list at the end of this lecture.

[1.2.4] Assume that $f \in D_M(a,F)$.

1°. When $M = B(E)$, f is said to be <u>Fréchet differentiable at</u> a and we denote this fact by $f \in D(a,F)$.

2°. When $M = K_s(E)$, f is said to be <u>Hadamard differentiable at</u> a and we denote this fact by $f \in D_H(a,F)$.

3°. When M is the class of all single point subsets of E, f is said to be <u>Gâteaux</u> <u>differentiable</u> <u>at</u> a and we denote this fact by $f \in D_G(a,F)$.

In each case, <u>the derivative will be denoted by the same</u> $f'(a)$ <u>and will be called the</u> Fréchet, Hadamard or Gâteaux <u>derivative respectively</u>. However, <u>in most cases, the term</u> "Fréchet" <u>will be omitted</u>.

The following relations are obvious.

(1.2.5) $\quad D(a,F) \subset D_H(a,F) \subset D_G(a,F) \subset D(a,F;\to E)$.

Later, in §1.4, the inverse relations will be discussed.

(1.2.6) $\quad$ <u>When</u> $E \in NLS$, $f \in D(a,F)$ <u>if and only if</u>

$$\lim_{x \to 0} \|x\|^{-1} r(f,a,x) = 0.$$

PROOF. $\Rightarrow$. For any $\{x_n\} \subset E$ such that $\|x_n\| \to 0$, put $\varepsilon_n = \|x_n\|$. Then, $\{\varepsilon_n^{-1} x_n\} \in B(E)$ and, by the assumption,

$$\lim_{n \to \infty} \|x_n\|^{-1} r(f,a,x_n) = \lim_{n \to \infty} \varepsilon_n^{-1} r(f,a,\varepsilon_n(\varepsilon_n^{-1} x_n)) = 0$$

$\Leftarrow$. If $f \notin D(a,F)$, there are $\{\varepsilon_n\} \subset R$ and $\{x_n\} \in B(E)$ such that $\varepsilon_n \to 0$ and $\varepsilon_n^{-1} r(f,a,\varepsilon_n x_n) \not\to 0$. Put $y_n = \varepsilon_n x_n$. Then, $y_n \to 0$ and

$$\|y_n\|^{-1} r(f,a,y_n) = \|x_n\|^{-1} \varepsilon_n^{-1} r(f,a,\varepsilon_n x_n) \to 0,$$

a contradiction.

(1.2.7) $\quad f \in D_H(a,F)$ <u>if and only if there exists</u> $u \in L(E,F)$ <u>such that the following condition is satisfied</u> : <u>if</u> $s \in C([0,1],A)$, $s(0) = a$ <u>and the limit</u>

$$s'(0) = \lim_{\varepsilon \to 0^+} \varepsilon^{-1} [s(\varepsilon) - s(0)]$$

<u>exists, then</u>

$$u(s'(0)) = \lim_{\varepsilon \to 0^+} \varepsilon^{-1} [f(s(\varepsilon)) - f(s(0))].$$

PROOF. $\Rightarrow$. Since $\{\varepsilon^{-1}[s(\varepsilon) - s(0)], s'(0) : \varepsilon \in R^+\} \in K_s(E)$,

$$\varepsilon^{-1} r(f,a,s(\varepsilon) - s(0)) = \varepsilon^{-1} r(f,a,\varepsilon(\varepsilon^{-1}[s(\varepsilon) - s(0)])) \to 0$$

if $\varepsilon \to 0$. Thus, the following limit exists :

$$\lim_{\varepsilon \to 0^+} \varepsilon^{-1}[f(s(\varepsilon)) - f(s(0))] = f'(a)(s'(0)).$$

$\Leftarrow$. If $f \notin D_H(a,F)$, then $g = f - u \notin D_H(a,F)$ and g again satisfies the condition of the theorem with $u = 0$. Then, for the corresponding remainder, we have that there exist $V \in N(F)$, $\{\varepsilon_n\} \in (c_0)$ and $\{x_n\} \in K_s(E)$ such that $\varepsilon_n^{-1} r(g,a,\varepsilon_n x_n) \notin V$. Construct $r \in C([0,1],A)$ such that

$$r(\varepsilon_n) = a + \varepsilon_n x_n, \quad r(0) = a \quad \text{and} \quad r'(0) = 0.$$

(see the remark below). Then,

$$\varepsilon_n^{-1} r(g,a,\varepsilon_n x_n) = \varepsilon_n^{-1}[g(a+\varepsilon_n x_n) - g(a)]$$

$$= \varepsilon_n^{-1}[g(r(\varepsilon_n)) - g(r(0))] \to 0 \quad \text{if } n \to \infty,$$

a contradiction.

REMARK. If $x_n \to x_0$, $a \in E$, $\varepsilon_n \downarrow 0$ and $\varepsilon_1 = 1$, then there exists $s \in C([0,1),E)$ such that $s(0) = a$, $s(\varepsilon_n) = a + \varepsilon_n x_n$ and $s'(0) = x_0$.

PROOF. Put

$$s_1(\varepsilon) = \begin{cases} \dfrac{\varepsilon_n - \varepsilon}{\varepsilon_n - \varepsilon_{n+1}} \varepsilon_{n+1} x_{n+1} + \dfrac{\varepsilon - \varepsilon_{n+1}}{\varepsilon_n - \varepsilon_{n+1}} \varepsilon_n x_n \\ \quad \text{if } \varepsilon_{n+1} \leq \varepsilon \leq \varepsilon_n; \\ 0 \quad \text{if } \varepsilon = 0. \end{cases}$$

Then, $s(\varepsilon) = s_1(\varepsilon) + a$ is the required mapping.

A mapping which satisfies the condition in the above theorem has been called a quasi-differentiable mapping by Dieudonné [1, p.151].

Next, we consider the so-called "chain rule" of the first order.

[1.2.8] An M-differentiation is said to have <u>the composition property</u> if $f \in D_M(a,F)$ and $g \in D_M(f(a),G)$, where $g : A_1 \to G \in TLS$ and $f(a) \in A_1 \in \mathcal{O}(F)$, then $g \circ f \in D_M(a,G)$ and $(g \circ f)'(a) = g'(f(a)) \circ f'(a)$.

(1.2.9) 1°. <u>Fréchet differentiation has the composition property</u>.

2°. <u>Hadamard differentiation has the composition property</u>.

3°. <u>Gâteaux differentiation does not have the composition property</u>.

PROOF. Generally, if f and g are as in (1.2.8), then

$$(g \circ f)(a+\varepsilon x) - (g \circ f)(a)$$
$$= g'(f(a))(f'(a)(\varepsilon x)) + g'(f(a))(r(f,a,\varepsilon x)) + r(g,f(a),f(a+\varepsilon x)-f(a)).$$

1°. Assume that $g'(f(a)) \circ f'(a)$ is not the Fréchet derivative of $g \circ f$. There exist $\{\varepsilon_n\} \in c_0$ and $\{x_n\} \in \mathcal{B}(E)$ such that

$$\varepsilon_n^{-1}[g'(f(a))(r(f,a,\varepsilon_n x_n)) + r(g,f(a),f(a+\varepsilon_n x_n) - f(a))] \not\to 0.$$

However, since $f \in D(a,F)$ and $g'(f(a)) \in L(F,G)$,

$$\varepsilon_n^{-1} g'(f(a))(r(f,a,\varepsilon_n x_n)) \to 0,$$

and since

$$\{\varepsilon_n^{-1}[f(a+\varepsilon_n x_n) - f(a)]\} = \{f'(a)(x_n) + \varepsilon_n^{-1} r(f,a,\varepsilon_n x_n)\} \in \mathcal{B}(F),$$

we have

$$\varepsilon_n^{-1} r(g,f(a),f(a+\varepsilon_n x_n) - f(a)) \to 0.$$

Thus, $g \circ f \in D(a,G)$ and $(g \circ f)'(a) = g'(f(a)) \circ f'(a)$.

2°. Similar as above.

3°. Define $f : R \to R^2$ and $g : R^2 \to R$ by $f(\xi) = (\xi \cos \xi, \xi \sin \xi)$ and $g(r,\theta) = r^2/\theta^3$ if $0 < \theta < 2\pi$; $= 0$ if $\theta = 0$. Then, $f \in D(0,R^2)$, $g \in D_G(0,R)$ and $g \circ f(0) = 0$. However, $g \circ f(\xi) = 1/\xi$. Thus, $g \circ f \notin D_G(0,R)$.

On the other hand, define $f : R \to R^2$ and $g : R^2 \to R$ by $f(\xi) = (\xi, \xi^2)$ and $g(\xi, \eta) = \xi$ if $\xi = \eta^2$; $= 0$ otherwise. Then, $f \in D(0, R^2)$ and $g \in D_G(0, R)$, and, moreover, $g \circ f = 1$. However, since $f'(0)(\xi) = (\xi, 0)$ and $g'(0) = 0$, we do not have the composition formula.

The following propositions will be used later.

(1.2.10) <u>Assume that</u> $f \in D_H(a, F)$, $F \in LCS$ <u>and</u> $\bar{a} \in \bar{F}$. <u>Then, for fixed</u> $x, y \in E$, <u>the mapping</u> $\phi : R^2 \to R$ <u>defined by</u>

$$\phi(\xi, \eta) = \langle f(a+\xi x+\eta y), \bar{a} \rangle$$

<u>is totally differentiable at</u> $(\xi, \eta) = (0, 0)$ <u>and</u>

$$\left(\frac{\partial \phi}{\partial \xi}\right)_{\xi=\eta=0} = \langle f'(a)(x), \bar{a} \rangle \quad \text{and} \quad \left(\frac{\partial \phi}{\partial \eta}\right)_{\xi=\eta=0} = \langle f'(a)(y), \bar{a} \rangle.$$

PROOF. We have to show that

$$\phi(\xi, \eta) - \phi(0, 0) = \xi \left(\frac{\partial \phi}{\partial \xi}\right)_{\xi=\eta=0} + \eta \left(\frac{\partial \phi}{\partial \eta}\right)_{\xi=\eta=0} + O(\|(\xi, \eta)\|),$$

where $\|(\xi, \eta)\| = (\xi^2 + \eta^2)^{1/2}$. By the assumption, the two partial derivatives exist and equal $\langle f'(a)(x), \bar{a} \rangle$ and $\langle f'(a)(y), \bar{a} \rangle$ respectively. Hence

$$\|(\xi, \eta)\|^{-1} [\phi(\xi, \eta) - \phi(0, 0) - \xi \left(\frac{\partial \phi}{\partial \xi}\right)_{\xi=\eta=0} - \eta \left(\frac{\partial \phi}{\partial \eta}\right)_{\xi=\eta=0}]$$

$$= \langle \|(\xi, \eta)\|^{-1} r(f, a, \|(\xi, \eta)\| (\|(\xi, \eta)\|^{-1} (\xi x+\eta y))), \bar{a} \rangle \to 0$$

if $\|(\xi, \eta)\| \to 0$, because the set

$$\{\|(\xi, \eta)\|^{-1} (\xi x+\eta y) : (\xi, \eta) \neq (0, 0), |\xi|, |\eta| \leq 1\}$$

is sequentially compact.

(1.2.11) <u>Let</u> $E_i, F_i \in TLS$, $a_i \in A_i \in \mathcal{O}(E_i)$ <u>and</u> $f_i \in D_H(a_i, F_i)$ $(i=1, 2)$. <u>Then the mapping</u> $f_1 \times f_2 : A_1 \times A_2 \to F_1 \times F_2$, <u>defined by</u>

$$(f_1 \times f_2)(x_1, x_2) = (f_1(x_1), f_2(x_2)) \quad \text{for} \quad x_i \in A_i \quad (i=1, 2),$$

<u>satisfies</u> $f_1 \times f_2 \in D_H((a_1,a_2), F_1 \times F_2)$ and

$$(f_1 \times f_2)'(a_1,a_2) = f_1'(a_1) \times f_2'(a_2).$$

<u>The same statement holds for Fréchet differentiability.</u>

PROOF   is easy and omitted.

NOTE :   The idea of defining differentiability by uniform convergence on a class of bounded sets is due to Sebastião e Silva [1,2,3], Keller [1] and Sova [1,2]. The most detailed exposition can be found in Averbukh-Smolyanov [1,2].

Fréchet differentiation is so called because, as was shown in (1.2.6), it coincides with Fréchet's original definition (Fréchet [1]) when the spaces involved are normed. For further informations about such differentiations, see §1.12.

Hadamard differentiation was introduced by Sova [1] under the name of compact differentiation. Hadamard [1] has emphasized the importance of the idea of quasi-differentiability in the case of finite-dimensional spaces. The generalization of his idea to infinite-dimensional spaces has been discussed by Fréchet [2], Ky Fan [1] and de Foglio [1]. The equivalence of these two notions, i.e., theorem (1.2.7) was first observed by Sova [1] and then by Averbukh-Smolyanov [2].

Gâteaux [1] gave the definition of directional differentiability when E is simply a linear space and $F = R$. Afterwards, Lévy [1] has imposed the restriction that the derivative should be linear.

§1.3 <u>Mean value theorems</u>.

There are various forms of the mean value theorem. We start with the simplest case.

(1.3.1)   <u>Assume that</u> $\Lambda \subset [\alpha,\beta]$ <u>is countable</u>, $F \in$ LCS, $s \in C([\alpha,\beta],F)$ <u>and</u> $s \in D([\alpha,\beta] \setminus \Lambda, F)$. <u>Then</u>,

1°. <u>For any</u> $\bar{x} \in \bar{F}$ <u>there exists</u> $\theta \in (0,1)$ <u>such that</u>

$$< s(\beta) - s(\alpha), \bar{x} > = (\beta-\alpha) < s'(\alpha+\theta(\beta-\alpha)), \bar{x} >.$$

2°. $\quad s(\beta) - s(\alpha) \in (\beta-\alpha)|\overline{Co}|\{s'(\xi) : \xi \in [\alpha,\beta] \setminus \Lambda\}$.

PROOF. 1°. Apply the one-dimensional mean value theorem to the function $\xi \mapsto < s(\xi), \bar{x} >$.

2°. This follows immediately from 1° and the BIPOLAR THEOREM.

(1.3.2) 1°. <u>If</u> F <u>is not</u> LCS, <u>then the above conclusion is not necessarily true</u>.

2°. $|\overline{Co}|$ <u>in the above theorem can not be replaced by</u> $|Co|$.

PROOF. 1°. Consider $L^{\frac{1}{2}}(-\infty,\infty)$ with the metric

$$\|x-y\| = \int_{-\infty}^{\infty} |x(\xi)-y(\xi)|^{\frac{1}{2}} d\xi .$$

Consider $s : R \to L^{\frac{1}{2}}(-\infty,\infty)$ defined by

$$s(\tau)(\xi) = 1 \text{ if } 0 \leq \xi \leq \tau; = 0 \text{ otherwise.}$$

Then, s is differentiable and $s'(\tau) = 0$ for any $\tau \in R$. Thus,

$$s(1) - s(0) \notin |\overline{Co}|\{s'(\tau) : 0 \leq \tau \leq 1\} .$$

2°. In ℓ^2, put $e_n = (0,\ldots 0,1,0,\ldots)$ with 1 at the n-th place, and define $s : [0,1] \to \ell^2$ by

$$s(\xi) = \sum_{i=1}^{n+1} \frac{1}{2^{i-1}} e_i + (\xi + \frac{1}{2^n} - 1)e_{n+2} \text{ if } 1 - \frac{1}{2^n} \leq \xi < 1 - \frac{1}{2^{n-1}} ,$$

$$s(1) = \sum_{i=1}^{\infty} \frac{1}{2^{i-1}} e_i .$$

Then, s satisfies the conditions of (1.3.1), and, for

$$M = \{s'(\xi) : \xi \in [0,1], \xi \neq 1 - \frac{1}{2^n}, n = 0,1,2,\ldots\} ,$$

we have

$$s(1) - s(0) \notin |Co|(M) .$$

Thus, it should be remembered that the mean value theorem can only be used in LCS.

(1.3.3) Assume that $E \in TLS$, $F \in LCS$, $[a,a+x] \subset A \in \mathcal{O}(E)$ and $f \in D(A,F;\to E)$.

Then, 1°. For any $\tilde{x} \in \bar{F}$ there exists $\theta \in (0,1)$ such that

$$< f(a+x) - f(a), \bar{x} > = < f'(a+\theta x, x), \bar{x} >.$$

2°. For any $q \in P(F)$ there exists $\theta \in (0,1)$ such that

$$q(f(a+x) - f(a)) \leq q(f'(a+\theta x, x)).$$

3° $f(a+x) - f(a) \in |\overline{Co}|\{f'(y,x) : y \in [a,a+x]\}$.

4°. If $E,F \in NLS$ and $f \in D(A,F)$, then

$$\|f(a+x) - f(a)\| \leq \|x\| \sup_{y \in [a,a+x]} \|f'(y)\|.$$

PROOF. 1°. Apply (1.3.1). 1° to $s(\xi) = f(a+\xi x)$.

2°. Apply the Hahn-Banach theorem to 1°.

3°. Apply (1.3.1). 2° to $s(\xi) = f(a+\xi x)$.

4°. Apply 2° above to the case when q is a norm.

An immediate consequence is the following proposition.

(1.3.4) Assume that $E \in TLS$, $F \in LCS$, $A \in \mathcal{O}(E)$ is convex and

$$f \in C(A,F) \cap D(A,F;\to E).$$

Then, 1°. $f'(x,y) = 0$ for any $(x,y) \in A \times E$ if and only if f is constant on A.

2°. Assume that $A = E$. $f'(x,y)$ does not depend on $x \in E$ and $f(0) = 0$ if and only if $f \in L(E,F)$.

PROOF is easy and omitted.

The following property will be used in §1.9.

(1.3.5) **Assume that** $E \in TLS$, $F \in LCS$ **and** $A \in \mathcal{O}(E)$. **If** $f \in D_G(A,F)$ **and, for some convex set** $M \subset A$, $f'(M)$ **is a bounded set in** $L(E,F)$, **then, for any** $V \in \mathcal{N}(F)$ **and** $B \in \mathcal{B}(E)$, **there exists** $\alpha \in (0,1]$ **such that**

$$f(x) - f(y) \in V \text{ if } x - y \in \alpha B \text{ and } x,y \in M.$$

PROOF. Since $f'(M) \in \mathcal{B}(L(E,F))$, there exists $\alpha \in (0,1]$ such that the closure of $\alpha f'(M)(B)$ is contained in V. Thus, by (1.3.3).3°, if $x,y \in M$ and $x - y \in \alpha B$,

$$f(x) - f(y) \in |\overline{Co}|\{f'(y+\theta(x-y))(x-y) : 0 \le \theta \le 1\} \subset V.$$

As a special case, we have the following theorem.

(1.3.6) **Let** $E, F \in NLS$, A **be an open ball**, $f \in D_G(A,F)$ **and, for some convex set** $M \subset A$, $f'(M) \in \mathcal{B}(L(E,F))$. **Then,** f **is uniformly continuous on** M.

The following theorem is also an immediate consequence of the mean value theorem.

(1.3.7) **Assume that** $E \in TLS$, $F \in LCS$, $a \in A \in \mathcal{O}(E)$ **and** $f \in D(A \setminus \{a\}, F)$. **If there exists** $u \in L(E,F)$ **such that**

$$\lim_{x \to a, x \in A} f'(x) = u,$$

then, $f \in D(a,F)$ **and** $u = f'(a)$.

PROOF. For any $q \in P(F)$ and $B \in \mathcal{B}(E)$, by (1.3.3).2°, if $x \in B$,

$$q(\varepsilon^{-1}[f(a+\varepsilon x) - f(a) - u(\varepsilon x)])$$

$$\le q((f'(a+\theta\varepsilon x)-u)(x)) \le q_B(f'(a+\theta\varepsilon x)-u) \to 0 \text{ if } \varepsilon \to 0.$$

NOTE: Dieudonné [1, p.142] has stated that "the real nature of the mean value theorem is exhibited by writing it as an inequality, and not as an equality", because in the conventional form of the theorem : $f(b) - f(a) = f'(c)(b-a)$ in the one-dimensional case, "all one need to know is that $f'(c)$ is a number which lies between the g.ℓ.b. and ℓ.u.b. of f' in the interval [a,b]".

We have started with the conventional equality form (1.3.1). 1°, because this is the simplest method to obtain the inequality form (1.3.1). 2° and those in (1.3.3). As the example (1.3.2). 1° shows, local convexity is not a heavy restriction when the mean value theorem is considered.

The mean value theorems of the forms (1.3.1). 2° and (1.3.3). 3° were first obtained by Sova [1]. The examples in (1.3.2) are due to Averbukh-Smolyanov [1].

(1.3.7) has been transplanted from L. Schwartz [1].

§1.4 Relations among M-differentiabilities

Starting with the directionally differentiable mappings, we shall give sufficient conditions under which a given mapping becomes differentiable in a stronger sense.

Throughout this section, we assume that

$$E \in TLS, \ F \in LCS, \ a \in A \in \mathcal{O}(E) \text{ and } f : A \to F.$$

(1.4.1) Assume that $f \in D(A,F;\to E)$. If

1°. $f'(x,y)$ is weakly continuous at $a \in A$ with respect to x when y is fixed;

2°. $f'(a,y)$ is continuous at 0 with respect to y,
then $f \in D_G(a,F)$.

PROOF. We have only to show that $f'(a,y)$ is linear with respect to y, because the continuity on y then follows from 2°. Now, by (1.3.1). 1°, for any $\bar{x} \in \bar{F}$,

$$< f'(a,y+z) - f'(a,y), \bar{x} >$$

$$= \lim_{\varepsilon \to 0} \varepsilon^{-1} < f(a+\varepsilon y+\varepsilon z) - f(a+\varepsilon y), \bar{x} >$$

$$= \lim_{\varepsilon \to 0} \varepsilon^{-1} < f'(a+\varepsilon y+\varepsilon \theta z, \varepsilon z), \bar{x} > \ = \ < f'(a,z), \bar{x} >$$

In other words, separate continuity implies Gâteaux differentiability. Naturally, joint continuity implies stronger conclusion.

(1.4.2) <u>Assume that</u> $f \in D(A,F;\to E)$. <u>If</u> $f'(x,y)$ <u>is jointly continuous at</u> (a,y) <u>for any</u> $y \in E$, <u>then</u> $f \in D_H(a,F)$.

PROOF. Assume that $f \notin D_H(a,F)$. Then, there exist $\{\varepsilon_n\} \in c_0$ and $\{x_n\} \subset E$ such that $x_n \to x_0$ and $\varepsilon_n^{-1}[f(a+\varepsilon_n x_n)-f(a)-f'(a)(\varepsilon_n x_n)] \not\to 0$, where $f'(a)$ is the Gâteaux derivative which exists by (1.4.1). Then, by (1.3.3). 2°, for any $q \in P(F)$,

$$q(\varepsilon_n^{-1}[f(a+\varepsilon_n x_n) - f(a) - f'(a)(\varepsilon_n x_n)])$$

$$\leq q(f'(a+\theta_n \varepsilon_n x_n, x_n) - f'(a)(x_n)) \to 0 \text{ if } n \to \infty.$$

In order to get a condition for Fréchet differentiability, we need a stronger continuity of the Gâteaux derivative.

[1.4.3] Let $f \in D_G(A,F)$. If the mapping $f' : A \to L(E,F)$ is continuous at a, then f is said to <u>be</u> C^1 <u>at</u> a, and we denote this fact by $f \in C^1(a,F)$. If $f \in C^1(a,F)$ for any $a \in A$, we say that f <u>is a</u> C^1-<u>mapping</u> of A <u>into</u> F and denote this fact by $f \in C^1(A,F)$.

(1.4.4) $C^1(a,F) \subset D(a,F)$.

PROOF. By the assumption, the Gâteaux derivative $f'(x)$ exists for every $x \in A$. By (1.3.3). 2°, if $q \in P(F)$ and $x \in B \in \mathcal{B}(E)$,

$$q(\varepsilon^{-1} r(f,a,\varepsilon x)) \leq q((f'(a+\theta \varepsilon x) - f'(a))(x))$$

$$\leq q_B(f'(a+\theta \varepsilon x) - f'(a)) \to 0 \text{ if } \varepsilon \to 0.$$

The following theorem gives a convenient criterion for Hadamard differentiability.

(1.4.5) <u>Let</u> $E \in$ LCS. <u>Assume that</u> $f \in D_G(a,F)$. <u>If</u> f <u>is lipschitzian on</u> A <u>in the following sense</u> : <u>for any</u> $q \in P(F)$ <u>there exists</u> $p \in P(E)$ <u>such that</u>

$$q(f(x) - f(y)) \leq p(x-y) \quad \text{if} \quad x,y \in A,$$

<u>then</u> $f \in D_H(a,F)$.

PROOF. Assuming that $f \notin D_H(a,F)$, we take $\{\varepsilon_n\}$ and $\{x_n\}$ as in the proof of (1.4.2). Then, since f is lipschitzian on A,

$$q(\varepsilon_n^{-1}[f(a+\varepsilon_n x_n) - f(a) - f'(a)(\varepsilon_n x_n)])$$
$$\leq q(\varepsilon_n^{-1}[f(a+\varepsilon_n x_0) - f(a) - f'(a)(\varepsilon_n x_0)])$$
$$+ p(x_n-x_0) + q(f'(a)(x_n-x_0)) \to 0 \quad \text{if} \quad n \to \infty.$$

Since a semi-norm is lipschitzian, we have the following.

(1.4.6) <u>If a semi-norm is Gâteaux differentiable at a point, it is Hadamard differentiable there.</u>

NOTE: Theorems (1.4.1) and (1.4.4) are very old; for instance, Graves [1] mentions them briefly. (1.4.5) was taken from Dieudonné [1. p.152, Problem 7)].

§1.5 <u>Differentiability in spaces with projective topologies</u>

In this section, we assume that E is a LCS whose topology is the projective topology determined by the family $\{E_\sigma, \pi_\sigma\}$.

In other words, for each σ, $\pi_\sigma : E \to E_\sigma$ is linear and the topology of E is the weakest topology by which every π_σ is continuous. It is easy to see that

$$P(E) = \{ \sum_{\text{finite}} P_\sigma \circ \pi_\sigma : P_\sigma \in P(E_\sigma) \} .$$

As usual, we assume that E is Hausdorff, or, equivalently, that for any $x \in E$ there exists σ such that $\pi_\sigma(x) \neq 0$.

(1.5.1) <u>Let</u> $F \in$ LCS, $a \in A \in \mathcal{O}(F)$ <u>and</u> $f : A \to E$. <u>Then,</u> $f \in D(a,F)$ <u>if and only if the following conditions are satisfied.</u>

1°. $\pi_\sigma \circ f \in D(a,E_\sigma)$ <u>for each</u> σ;

2°. $\bigcap_\sigma \pi_\sigma^{-1}[(\pi_\sigma \circ f)'(a)(x)] \neq \phi$ <u>for each</u> $x \in F$.

PROOF. If $f \in D(a,E)$, since $\pi_\sigma \in L(E,E_\sigma)$, it follows from (1.2.9). 1° that $\pi_\sigma \circ f \in D(a,E_\sigma)$ and $(\pi_\sigma \circ f)'(a) = \pi_\sigma \circ f'(a)$. Therefore,

$$f'(a)(x) \in \pi_\sigma^{-1}[(\pi_\sigma \circ f)'(a)(x)].$$

To prove the converse, we define a mapping $u : F \to E$ by corresponding to each $x \in F$ an arbitrary element

$$u(x) \in \bigcap_\sigma \pi_\sigma^{-1}[(\pi_\sigma \circ f)'(a)(x)].$$

Then, u is linear and continuous. For this $u \in L(F,E)$, put

$$r(f,a,x) = f(a+x) - f(a) - u(x).$$

Then, for any $p = \sum p_\sigma \circ \pi_\sigma$ and any $B \in \mathcal{B}(E)$,

$$\sup_{x \in B} p[\epsilon^{-1} r(f,a,\epsilon x)]$$

$$\leq \sum \sup_{x \in B} p_\sigma[\epsilon^{-1} r(\pi_\sigma \circ f, a, \epsilon x)] \to 0 \quad \text{if} \quad \epsilon \to 0.$$

Thus, $f \in D(a,F)$.

In this theorem, Fréchet differentiability can be replaced by M-differentiability. However, it fails for strong Fréchet differentiability (see [1.7.3]). For example,

$$f : R^\infty \to R^\infty, \quad (x_1,x_2,\ldots) \mapsto (x_1^2, x_2^2, \ldots)$$

is not strongly Fréchet differentiable at 0, but for

$$\pi_n(x_1,x_2,\ldots) = x_n,$$

$\pi_n \circ f$ is strongly Fréchet differentiable at 0 and $(\pi_n \circ f)'(0) = 0$ for all n.

To study the C^1-case, we recall the fact that the topology of $L(F,E)$ is the projective topology determined by the family $\{L(F,E_\sigma),\Pi_\sigma\}$, where $\Pi_\sigma(u) = \pi_\sigma \circ u$ for every $u \in L(F,E)$.

(1.5.2) <u>Let</u> $F \in$ LCS, $a \in A \in \mathcal{O}(F)$ <u>and</u> $f \in D(A,E)$. <u>Then</u>, f <u>is</u> C^1 <u>at</u> a <u>if and only if</u> $\pi_\sigma \circ f$ <u>is</u> C^1 <u>at</u> a <u>for each</u> σ.

PROOF. If f is C^1 at a, the mapping $f' : A \to L(F,E)$ is continuous at a, which is equivalent to $\pi_\sigma \circ f' : A \to L(F,E_\sigma)$ continuous at a for each σ (Schaefer [1], p.51). However, for $x \in A$,

$$(\pi_\sigma \circ f')(x) = \pi_\sigma \circ f'(x) = (\pi_\sigma \circ f)'(x).$$

§1.6 Differentiability in inductive limits of increasing sequences of subspaces

The theorems on differentiability in this section will be proved only in a restricted class of spaces endowed with the inductive topology - the "regular" inductive limit of an increasing family of subspaces. Namely, we consider $E \in$ LCS and $E_\sigma \in$ LCS, where the set $\{\sigma\}$ is directed, such that the following conditions are satisfied :

1. $\sigma < \sigma_1$ implies $E_\sigma \subset E_{\sigma_1}$;

2. $E = \cup E_\sigma$;

3. The topology of E is the strongest LC topology such that each of the imbeddings $i_\sigma : E_\sigma \to E$ is continuous.

4. The topology of E_σ coincides with the relative topology by E;

5. $B \in \mathcal{B}(E)$ if and only if $B \subset E_\sigma$ for some σ and $B \in \mathcal{B}(E_\sigma)$.

(1.6.1) <u>Let</u> E <u>be as above</u>, $F \in$ LCS <u>and</u> $a \in A \in \mathcal{O}(E)$. <u>Then</u>, $f \in D(a,F)$ <u>if and only if</u> $f \circ i_\sigma \in D(a,F)$ <u>for each</u> σ <u>such that</u> $a \in E_\sigma$.

PROOF. Since the necessity is obvious, we assume that

$$f \circ i_\sigma : A \cap E_\sigma \to F$$

is differentiable at a for each σ such that $a \in E_\sigma$. We take an arbitrary $x \in E$. Then, by conditions 1 and 2, there exists σ such that $a, x \in E_\sigma$. For this σ, we put

$$u(x) = (f \circ i_\sigma)'(a)(x).$$

Obviously, $u(x)$ does not depend on σ as long as $a, x \in E_\sigma$, and, hence, it only depends on a and x. Moreover, if $a \in E_\sigma$,

$$(u \circ i_\sigma)(x) = u(x) = (f \circ i_\sigma)'(a)(x) \quad \text{if } x \in E_\sigma,$$

which implies $u \in L(E,F)$. For this u, we put

$$r(f,a,x) = f(a+x) - f(a) - u(x),$$

and assume that $\{\varepsilon^{-1} r(f,a,\varepsilon x)\}$ does not converge to zero uniformly for $x \in B$ for some $B \in \mathcal{B}(E)$ as $\varepsilon \to 0$. Then, there exist $\{\varepsilon_n\} \in c_0$ and $\{x_n\} \subset B$ such that $\varepsilon_n^{-1} r(f,a,\varepsilon_n x_n) \not\to 0$. By condition 5, we can find σ such that $a, x_n \in E_\sigma$ for all n. Then,

$$\varepsilon_n^{-1} r(f,a,\varepsilon_n x_n) = \varepsilon_n^{-1} r(f \circ i_\sigma, a, \varepsilon_n x_n) \to 0 ,$$

which is a contradiction.

To study the C^1-case, we recall that, under the conditions 1-5, the topology of $L(E,F)$ is the projective topology determined by the family $\{L(E_\sigma), \Pi_\sigma\}$, where $\Pi_\sigma(u) = u \circ i_\sigma$ for each σ (Schaefer [1], p.116).

(1.6.2) Let $a \in A \in \mathcal{O}(E)$, $F \in LCS$ and $f \in \mathcal{D}(A,F)$. Then, f is C^1 at a if and only if $f \circ i_\sigma$ is C^1 at a for each σ such that $a \in E_\sigma$.

PROOF. f is C^1 at a if and only if the mapping $f' : A \to L(E,F)$ is continuous at a, which is equivalent to

$$\pi_\sigma \circ f' : A \to L(E_\sigma, F)$$

continuous at a for each σ. On the other hand,

$$(\Pi_\sigma \circ f')(x) = f'(x) \circ i_\sigma = (f \circ i_\sigma)'(x) \quad \text{if} \quad x \in A \cap E_\sigma.$$

§1.7 Differentiability and continuity

It is not difficult to find a differentiable mapping which is not continuous. For instance, let E be a Hilbert space equipped with its weak topology. Then, $f(x) = \|x\|^2$ is differentiable, but it is not continuous. Further, the composition mapping is nowhere continuous as we shall see in Appendix 2, but can be infinitely continuously differentiable as we shall see in §3.1.

However, we do not need to look for such examples, because we have the following decisive theorem. For the definition and properties of sequential spaces, see Appendix 1.

(1.7.1) Let E,F $\in$ TLS and M be a system of bounded subsets of E which contains all convergent sequences. Then, $f \in D_M(a,F)$ implies that f is continuous at a if and only if E is sequential.

PROOF. Assume that $f \in D_M(a,F)$ and f is not continuous at a. Then, there exists a net $\{x_\lambda\} \subset E$ such that $x_\lambda \to a$ and $f(x_\lambda) \nrightarrow f(a)$. Taking a subnet if necessary, we can assume that $\{f(x_\lambda)\}$ is outside of a neighbourhood of f(a). Since E is sequential, we can choose a subsequence $\{x_n\} \subset \{x_\lambda\}$ such that $x_n \to a$. For the M-derivative $f'(a)$, we have

$$f(x_n) - f(a) = f'(a)(x_n - a) + r(f, a, x_n - a).$$

Put $y_n = x_n - a$ and choose a subsequence $\{y_{n_i}\}$ such that $n_i y_{n_i} \to 0$ if $i \to \infty$. Thus,

$$r(f, a, x_{n_i} - a) = r(f, a, n_i^{-1}(n_i y_{n_i})) \to 0 \quad \text{if} \quad i \to \infty,$$

which implies $f(x_{n_i}) \to f(a)$. This contradicts $f(x_{n_i})$ outside some neighbourhood of f(a).

To prove the converse, assume that E is not sequential. Then, there is a set $M \subset E$ and a limit point a such that a is not the limit of any sequence taken from M. Let f be the characteristic function of M. Then, f is not continuous at a. However, $f \in D_M(a,F)$ for $M = B(E)$, because, for any $\{\varepsilon_n\} \in c_0$ and $\{x_n\} \in B(E)$, since $a + \varepsilon_n x_n \to a$, $\{a+\varepsilon_n x_n\}$ is not contained in M except for finitely many n. Thus,

$$\varepsilon_n^{-1}[f(a+\varepsilon_n x_n) - f(a)] \to 0 \quad \text{if} \quad n \to \infty.$$

Therefore, if $E \in$ MLS and $f \in D_H(a,F)$, then f is continuous at a. On the other hand, even if $f \in D(A,F)$ and $a \in A \in O(E)$, f is not necessarily continuous at a, as the above example of Keller shows.

Following Floret-Wloka [1], we shall call $E \in$ LCS an (LS)-space if it is an inductive limit of a sequence $\{E_n\} \subset$ LCS such that the embeddings $E_n \to E_m$ are strongly compact (for the definition, see §2.1) whenever $m > n$. In this case, the five conditions in §1.6 are all satisfied, and, moreover, for any topological space F, $f : E \to F$ is continuous if and only if $f/E_n : E_n \to F$ are continuous for all n. Further, we can assume that $E_n \in$ BS.

Therefore the following theorem is now obvious.

(1.7.2) <u>Let</u> E <u>be an</u> (LS)-<u>space and</u> $F \in$ TLS. <u>Then, if</u> $f \in D(a,F)$ <u>for some</u> $a \in E$, f <u>is continuous at</u> a.

It is of course possible to define a differentiability which implies continuity. Several such definitions will be discussed in §1.12. Here, we shall introduce the strongest one of them.

[1.7.3] Let $E,F \in$ TLS, $a \in A \in O(E)$ and $f : A \to F$. f is said to be <u>strongly (Fréchet) differentiable at</u> a if $f \in D_G(a,F)$ and there exists $U \in N(E)$ such that

$$\lim_{\varepsilon \to 0} \varepsilon^{-1} r(f,a,\varepsilon x) = 0 \quad \text{uniformly for} \quad x \in U.$$

The proof of the following fact is now obvious.

(1.7.4)   If f is strongly differentiable at a, then f ∈ D(a,F) and is continuous at a.

NOTE: (1.7.1) was proved by Balanzat [2] and refined by Averbukh-Smolyanov [2]. The earlier paper of Keller [1] contains a proof of the fact that if E ∈ MLS and f is differentiable at a point then it is continuous there.

(1.7.2) is due to Averbukh-Smolyanov [2].   The definition (1.7.3) seems to have been given first by Keller [1].

§1.8 Higher derivatives

It is quite simple to define higher M-derivatives in general TLS, but the properties of these higher derivatives are far from simple. In fact, we shall meet here one of the biggest differences between the normed and the non-normed cases.

We assume that

$$E, F \in \text{TLS}, \ a \in A \in \mathcal{O}(E) \ \text{ and } \ f : A \to F.$$

[1.8.1]   f is said to be twice M-differentiable at a if $f \in D_M(A,F)$ and the mapping $f' : A \to L(E,F)$ is M-differentiable at a. We denote this fact by $f \in D_M^2(a,F)$. The second M-derivative of f at a is denoted by f"(a) or $f^{(2)}(a)$, which is, by definition, an element of $L(E,L(E,F))$, which we shall denote by $L^2(E,F)$.

Generally, f is said to be n-times M-differentiable at a if $f \in D_M^{n-1}(A,F)$ and the mapping $f^{(n-1)} : A \to L^{n-1}(E,F)$ is M-differentiable at a. We denote this fact by $f \in D_M^n(a,F)$.

The notion of C^n-mapping can be defined in the same way.

The space $L^n(E,F)$ is different from $L(E^n,F)$ that is the space of all continuous multilinear mappings of $E^n = E \times \ldots \times E$ (n-times) to F.

We shall denote the value of the derivatives in the following way :

$$f"(a)(x_2)(x_1) = \lim_{\varepsilon \to 0} \varepsilon^{-1}[f'(a+\varepsilon x_2)(x_1) - f'(a)(x_1)]$$

and

$$f^{(n)}(a)(x_n)(x_{n-1})\ldots(x_1) = \lim_{\varepsilon \to 0} \varepsilon^{-1}[f^{(n-1)}(a+\varepsilon x_n)(x_{n-1})\ldots(x_1)$$
$$- f^{(n-1)}(a)(x_{n-1})\ldots(x_1)].$$

Moreover, we put

$$f^{(n)}(a)(x_1,x_2,\ldots,x_n) = f^{(n)}(a)(x_n)\ldots(x_1).$$

(1.8.2) <u>If</u> $F \in LCS$ <u>and</u> $f \in D_H^n(a,F)$, <u>then</u> $f^{(n)}(a)$ <u>is symmetric</u>.

PROOF. We shall prove only the case when $n = 2$. Take $\bar{a} \in \bar{F}$ and consider the function $\phi : R^2 \to R$ defined by

$$\phi(\xi,\eta) = < f(a+\xi x+\eta y), \bar{a} >,$$

where x and y are fixed. Obviously, ϕ is defined in a neighbourhood of $(0,0)$. By (12.10), $\frac{\partial \phi}{\partial \xi}$ and $\frac{\partial \phi}{\partial \eta}$ exist and are totally differentiable at $(0,0)$. Thus by Young's theorem (Young [1]),

$$\frac{\partial}{\partial \eta}(\frac{\partial \phi}{\partial \xi}) = \frac{\partial}{\partial \xi}(\frac{\partial \phi}{\partial \eta}) \quad \text{at} \quad (0,0),$$

which means

$$< f^{(2)}(a)(x,y), \bar{a} > = < f^{(2)}(a)(y,x), \bar{a} >.$$

Since $\bar{a}$ is arbitrary, we have the conclusion.

As we have seen in (1.2.9), Fréchet and Hadamard differentiations satisfy the first order chain rule. They satisfy the higher order chain rules also.

(1.8.3) <u>Let</u> M <u>be either</u> $B(E)$ <u>or</u> $K_s(E)$. <u>If</u> $f \in D_M^n(A,F)$ <u>and</u> $g \in D_M^n(A_1,G)$, <u>where</u> $f(A) \subset A_1 \in \mathcal{O}(F)$ <u>and</u> $G \in TLS$, <u>then</u> $g \circ f \in D_M^n(A,G)$ <u>and</u>

$$(g \circ f)^{(n)}(a)(x_1,x_2,\ldots,x_n)$$
$$= \sum_{k=1}^{n} \sum g^{(k)}(f(a))(f^{(j_1)}(a)(x_{1,j_1},\ldots,x_{j_1,j_1}),\ldots,f^{(j_k)}(a)(x_{1,j_k},\ldots,x_{j_k,j_k})),$$

where the second $\sum$ is taken over all partitions of the set $\{x_1,\ldots,x_n\}$ into k disjoint sets

$$\{x_{1,j_1},\ldots,x_{j_1,j_1}\},\ldots, \{x_{1,j_k},\ldots,x_{j_k,j_k}\}$$

such that each of these sets is in the increasing order of the indices and $j_1 + \ldots + j_k = n$.

PROOF. The case when $n = 1$ has been proved in (1.2.9). The general case should, of course, be proved by induction. But, to avoid the complicated notations, we shall only prove the case when $n = 2$. Let us consider the bilinear mapping $u : E^2 \to F$ defined by

$$u(x)(y) = g''(f(a))(f'(a)(x)(f'(a)(y) + g'(f(a))(f''(a)(x)(y)).$$

Then, clearly $u(x) \in L(E,F)$ for each $x \in E$ and, moreover, $u \in L^2(E,F)$. To show that u is the second derivative of $g \circ f$, let $\{\varepsilon_n\} \in c_0$ and $\{x_n\} \in B(E)$. Then,

$$\varepsilon_n^{-1}[(g \circ f)'(a+\varepsilon_n x_n) - (g \circ f)'(a) - (g''(f(a))(f'(a)(\varepsilon_n x_n)) \circ f'(a)$$
$$- g'(f(a)) \circ f''(a)(\varepsilon_n x_n)]$$
$$= \{g''(f(a))(\varepsilon_n^{-1}r(f,a,\varepsilon_n x_n)) \circ f'(a) + (\varepsilon_n^{-1}r(g',f(a),f(a+\varepsilon_n x_n) - f(a))) \circ f'(a)$$
$$+ g'(f(a)) \circ (\varepsilon_n^{-1}r(f',a,\varepsilon_n x_n))\}$$
$$+ \varepsilon_n\{g''(f(a))(f'(a)(x_n)) \circ f'(a) + g''(f(a))(f'(a)(x_n)) \circ (\varepsilon_n^{-1}r(f',a,\varepsilon_n x_n))$$
$$+ g''(f(a))(\varepsilon_n^{-1}r(f,a,\varepsilon_n x_n)) \circ (\varepsilon_n^{-1}r(f',a,\varepsilon_n x_n))$$
$$+ (\varepsilon_n^{-1}r(g',f(a),f(a+\varepsilon_n x_n) - f(a))) \circ (\varepsilon_n^{-1}r(f',a,\varepsilon_n x_n))\}.$$

Each member of the first group converges to zero because of the separate continuity of the composition mapping. Each member of the second group is a bounded sequence in $L(E,G)$ and, hence, being multiplied by ε_n, they converge to zero.

The following is Taylor's theorem. We put

$$f^{(n)}(a)(x,\ldots,x) = f^{(n)}(a)(x^n).$$

(1.8.4) **Let** $F \in \text{LCS}$, $f \in D_G^n(A,F)$ **and** $[a,a+x] \subset A$. **Then, for any** $q \in P(F)$ **there exists** $\theta \in (0,1)$ **such that**

$$q(f(a+x) - \sum_{i=0}^{n-1} \frac{1}{i!} f^{(i)}(a)(x^i)) \leq \frac{1}{n!} q(f^{(n)}(a+\theta x)(x^n)).$$

If, moreover, $E,F \in \text{NLS}$, **we have**

$$\|f(a+x) - \sum_{i=0}^{n-1} \frac{1}{i!} f^{(i)}(a)(x^i)\| \leq \frac{1}{n!} \|x\|^n \sup_{z \in [a,a+x]} \|f^{(n)}(z)\|.$$

PROOF. For $\bar{a} \in \bar{F}$, consider the function $\phi : R \to R$ defined by

$$\phi(\xi) = \langle f(a+\xi x), \bar{a} \rangle$$

and apply the one-dimensional Taylor's theorem.

As the converse of Taylor's theorem, Abraham-Robbin [1] gives the following theorem. For the proof, we refer to page 6 of [1].

(1.8.5) **Let** $E,F \in \text{BS}$, $A \in \mathcal{O}(E)$ **be convex. If there exist** u_i **of** A **into the spaces of symmetric linear mappings of** E^i **into** F **such that, for**

$$r(x,y) = f(x+y) - \sum_{i=0}^{n} \frac{1}{i!} u_i(x)(y^i),$$

1°. **each** u_i **is continuous**;

2°. $\|y\|^{-n} \|r(x,y)\| \to 0$ **if** $(x,y) \to (x_0,0)$,

then, $f \in C^n(A,F)$ **and** $f^{(i)} = u_i$ **for each** i.

NOTE: The composition mapping comp : $L(E,F) \times L(F,G) \to L(E,G)$ for $E,F,G \in \text{TLS}$ defined by $\text{comp}(u,v) = v \circ u$ is, in general, not continuous. In the proof of (1.8.3), the following fact was used.

"If $\tilde{B}_1 \in B(L(E,F))$ and $\tilde{B}_2 \in B(L(F,G))$, then $\tilde{B}_2 \circ \tilde{B}_1 \in B(L(E,G))$".

In other words, the mapping comp is bounded without being continuous. We shall discuss this in more detail in Appendix 2.

The fact that comp is not always continuous has caused many difficulties. For example,

(1) For C^1-mappings defined in [1.4.3], the first order chain rule may not hold. Let $f \in C^1(A,F)$ and $g \in C^1(A_1,G)$ using the same notions as in (1.8.3). Then, by (1.2.9) and (1.4.4), we have that $g \circ f \in D(A,F)$ and $(g \circ f)'(x) = g'(f(x)) \circ f'(x)$. As we have seen in §1.7, f in $g'(f(x))$ may not be continuous. However, even if it is continuous, $(g \circ f)'$ itself, being a composition, may not be continuous.

This difficulty does not occur if E is sequential, because of (1.7.1) and Appendix 2, (3.2).

(2) The strong Fréchet differentiability defined in (1.7.3) may not satisfy the second order chain rule, because, in the proof of (1.8.3), the members of the second group are no longer sequences.

These facts are the origin of the theory of calculus in spaces with limit structure or "Limesraüme", which has been investigated by Binz [1], Binz-Mier-Solfrian [1], Fisher [1], Fröhlicher-Bucher [1], Keller [2,3] and others. The pseudo-topological linear spaces introduced by Marinescu [2] are also of the same origin. On these spaces, which are not topological linear spaces, one can construct a theory of differentiation in which every differentiable mapping is continuous and the chain rules of all order hold.

In this lecture, we remain within the frame of TLS, in particular, LCS. So far, we have met two difficulties : differentiable mappings may not be continuous, and C^1-mappings may not satisfy the chain rule. In the next section, we shall introduce a new notion by which these difficulties can be avoided.

Kijowski-Szczyrba [1] and Szczyrba [1] have presented new methods to overcome these difficulties remaining within the frame of LCS.

§1.9 Equicontinuously differentiable mappings

The aim of this section is to introduce a notion of equicontinuous differentiability, which may be regarded as a TLS version of continuous differentiability in NLS.

In this section, we assume that

$$E, F \in TLS, \quad a \in A \in \mathcal{O}(E) \quad \text{and} \quad f : A \to F.$$

[1.9.1] f is said to be <u>equicontinuously differentiable at</u> a or $f \in EC^1(a,F)$ if $f \in D(A,F)$ and the mapping $f' : A \to L(E,F)$ satisfies the following conditions :

(EC) 1°. f' is continuous at a;

(EC) 2°. there exists $U_a \in N(E)$ such that $f'(a+U_a)$ is an equicontinuous subset of $L(E,F)$.

f is said to be <u>n-times equicontinuously differentiable</u> at a or $f \in EC^n(a,F)$ if $f \in D^n(A,F)$, $f \in EC^k(a,F)$ for $k < n$ and the mapping $f^{(n)} : A \to L^n(E,F)$ satisfies the above two conditions.

It is obvious that $EC^n(a,F) \subset C^n(a,F)$. As the converse, we have the following theorem.

(1.9.2) <u>If</u> $E, F \in NLS$, <u>every</u> C^n-<u>mapping is</u> n-<u>times equicontinuously differentiable</u>.

PROOF. Assume that, for some $k \leq n$, $f^{(k)}$ does not satisfy (EC) 2°. then, there exist $\{x_i\}$ and $\{z_i\}$ such that $x_i \to 0$, $z_i \to 0$ and $f^{(k)}(a+x_i)(z_i) \not\to 0$, which is a contradiction.

As we have mentioned in §1.7, there is a C^∞-mapping which is not continous. However,

(1.9.3) <u>If</u> $F \in LCS$ <u>and</u> $f \in EC^1(a,F)$, <u>then</u> f <u>is continuous at</u> a.

PROOF. For any $V \in N(F)$, we take $V_1 \in N(F)$ such that $\bar{V}_1 \subset V$. By (EC) $2°$, there exists $U \in N(E)$ such that

$$f'(a+U)(U) \subset V_1.$$

Then, by the mean value theorem (1.3.3) $3°$, if $x \in U$,

$$f(a+x) - f(a) \in \overline{|Co|}\{f'(a+\theta x)(x) : 0 \leq \theta \leq 1\} \subset \bar{V}_1 \subset V.$$

Now, we prove the chain rule for EC^n-mappings.

(1.9.4) <u>Let</u> $F,G \in LCS$, $f(a) \in A_1 \in \mathcal{O}(F)$ <u>and</u> $g : A_1 \to G$. <u>If</u> $f \in C^n(a,F)$ <u>and</u> $g \in EC^n(f(a),G)$, <u>then</u> $g \circ f \in EC^n(a,G)$.

PROOF. We shall prove only the cases $n = 1$ and $n = 2$. The case $n = 1$:

Let $x_\lambda \to 0$ in E. Since f is continuous and $g' : A_1 \to L(F,G)$ satisfies (EC) $2°$, there exists λ_0 such that

$$\{g'(f(a+x_\lambda)) : \lambda \geq \lambda_0\}$$

is equicontinuous. On the other hand,

$$(g \circ f)'(a+x_\lambda) - (g \circ f)'(a)$$

$$= g'(f(a+x_\lambda)) \circ [f'(a+x_\lambda)-f'(a)] + [g'(f(a+x_\lambda)) - g'(f(a))] \circ f'(a).$$

By (2.3) of Appendix 2, the first member converges to zero. By the separate continuity of comp, the second member also converges to zero. Thus, (EC) $1°$ is satisfied.

To show that (EC) $2°$ is satisfied, we take $U_a \in N(E)$ and $V_b \in N(F)$ such that $f'(a+U_a)$ and $g'(b+V_b)$ are equicontinuous subsets of $L(E,F)$ and $L(F,G)$ respectively, where $b = f(a)$. Since f is continuous, there exists $U_1 \in N(E)$ such that

$$f(a+U_1) - f(a) \subset V_b.$$

Let $W \in N(G)$. Then, there exist $U \in N(E)$ and $V \in N(F)$ such that

$$f'(a+U_a)(U) \subset V_0 \text{ and } g'(b+V_b)(V) \subset W.$$

Then,

$$(g \circ f)'(a+U_a \cap U_1)(U) \subset g'(f(a+U_1))(f'(a+U_a))(U) \subset g'(b+V_b)(V) \subset W,$$

which shows that $(g \circ f)'(a+U_a \cap U_1)$ is equicontinuous. The case $n = 2$:

Let $x_\lambda \to 0$ in E. We assume that $f'(a+U_a) \subset L(E,F)$, $f''(a+U_a) \subset L^2(E,F)$, $g'(b+V_b) \subset L(F,G)$ and $g''(b+V_b) \subset L^2(F,G)$ are all equicontinuous. Now, we have

$$(g \circ f)''(a+x_\lambda)(z) - (g \circ f)''(a)(z)$$

$$= g''(f(a+x_\lambda))(f'(a+x_\lambda)(z)) \circ [f'(a+x_\lambda) - f'(a)] \quad \ldots (1)$$

$$+ g''(f(a+x_\lambda))[f'(a+x_\lambda)(z) - f'(a)(z)] \circ f'(a) \quad \ldots (2)$$

$$+ [g''(f(a+x_\lambda)) - g''(f(a))](f'(a)(z)) \circ f'(a) \quad \ldots (3)$$

$$+ g'(f(a+x_\lambda)) \circ [f''(a+x_\lambda)(z) - f''(a)(z)] \quad \ldots (4)$$

$$+ [g'(f(a+x_\lambda)) - g'(f(a))] \circ f''(a)(z) \quad \ldots (5).$$

We take $B_0 \in B(E)$, $B \in B(E)$ and $W \in N(G)$. Then,

$$\tilde{W} = (B,W) \in N(L(E,G))$$

To prove that (EC) 1° is satisfied, we shall show that each of the above five terms satisfies the following condition: there exists λ_0 such that, if $\lambda \geq \lambda_0$, the term is contained in $\tilde{W}$ uniformly for $z \in B_0$.

re. (1). Since $B_1 = f'(a+U_a)(B_0) \in B(F)$, we have $(B_1,W) \in N(L(F,G))$. Since g'' satisfies (EC) 2° and f is continuous at a, there is λ_1 such that $\lambda \geq \lambda_1$ implies $f(a+x_\lambda) \in b + V_b$ and there exists $V \in N(F)$ such that $g''(f(a+x_\lambda))(V) \subset (B_1,W)$. Since f' is continuous, there is λ_2 such that $\lambda \geq \lambda_2$ implies $f'(a+x_\lambda) - f'(a) \in (B,V)$. Thus, for λ_0 such that $\lambda_0 \geq \lambda_1, \lambda_2$, $\lambda \geq \lambda_0$ implies

$$(1)(B) \subset g''(f(a+x_\lambda))(B_1)(V) = g''(f(a+x_\lambda))(V)(B_1) \subset W,$$

and hence $(1) \in W$.

re. (2). Consider the following set :

$$B_2 = f'(a)(B) \in \mathcal{B}(F) \quad \text{and} \quad (B_2, W) \in \mathcal{N}(L(F,G)).$$

Then, there is λ_0 such that $\lambda \geq \lambda_0$ implies $g''(f(a+x_\lambda))(V) \subset (B_2, W)$ and $f'(a+x_\lambda) - f'(a) \in (B_0, V)$. Hence, $\lambda \geq x_0$ implies

$$(2)(B) \subset g''(f(a+x_\lambda))[f'(a+x_\lambda)(z) - f'(a)(z)](B_2) \subset g''(f(a+x_\lambda))(V)(V_2) \subset W,$$

or $(2) \in \hat{W}$.

re. (3). Consider the following sets :

$$B_3 = f'(a)(B_0) \in \mathcal{B}(F) \quad \text{and} \quad (B_3, (B_2, W)) \in \mathcal{N}(L^2(F,G)).$$

Since g'' and f are continuous, there is λ_0 such that $\lambda \geq \lambda_0$ implies $g''(f(a+x_\lambda)) - g''(f(a)) \in (B_3, (B_2, W))$, and, hence,

$$(3)(B) \subset [g''(f(a+x_\lambda)) - g''(f(a))](B_3)(B_2) \subset W,$$

or $(3) \in \hat{W}$.

re. (4). There is λ_1 and $V \in \mathcal{N}(F)$ such that $\lambda \geq \lambda_1$ implies $g'(f(a+x_\lambda))(V) \subset W$. Since f'' is continuous, there is λ_2 such that $\lambda \geq \lambda_2$ implies $f''(a+x_\lambda) - f''(a) \in (B_0, (B,V))$. Thus, for λ_0 such that $\lambda_0 \geq \lambda_1, \lambda_2$, $\lambda \geq \lambda_0$ implies

$$(4)(B) \subset g'(f(a+x_\lambda))(V) \subset W,$$

or $(4) \in \hat{W}$.

re. (5). Since $f''(a)(B_0)$ is bounded, $f''(a)(B_0)(B)$ is bounded. Hence the conclusion follows from the continuity of f and g'.

To prove that (EC) 2° is satisfied, we use W defined above and take $W_1 \in \mathcal{N}(G)$ such that $W_1 + W_1 \subset W$. There are $U_0 \in \mathcal{N}(E)$ and $V \in \mathcal{N}(F)$ such that

$U_0 \subset U_a$ and

$$g''(f(a+U_0))(V) \subset (B_2,W_1) \quad \text{and} \quad g'(f(a+U_0))(V) \subset W_1.$$

Finally, we take $U \in N(E)$ such that

$$f'(a+U_0)(U) \subset V \quad \text{and} \quad f''(a+U_0)(U) \subset (B,V).$$

Then,

$$(g \circ f)''(a+U_0)(U)(B) \subset g''(f(a+U_0)(f'(a+U_0)(U))(f'(a)(B))$$

$$+ g'(f(a+U_0))(f''(a+U_a)(U)(B))$$

$$\subset g''(f(a+U_0))(V)(B_2) + g'(f(a+U_0))(V) \subset W_1 + W_1 \subset W,$$

or $(g \circ f)''(a+U_0)(U) \subset \widetilde{W}$.

Obviously, $L(E,F) \subset EC^\infty(E,F)$. In §4.3, we shall show that any continuously differentiable semi-norm is equicontinuously differentiable.

The proof of the following theorem is routine and omitted.

(1.9.5) <u>Under the same assumption as</u> (1.2.11) <u>except that the spaces are now assumed to be locally convex, if</u> $f_i \in EC^n(a_i,F_i)$ (i=1,2), <u>then</u> $f_1 \times f_2 \in EC^n((a_1,a_2),F_1 \times F_2)$.

NOTE: Lloyd [3] has considered the mappings which satisfy (EC) 1° and (EC) 2° where the equicontinuity is replaced by boundedness. In this case, the mapping may not be continuous, but the chain rules are satisfied. Penot [1] also has considered both cases independently.

§1.10 Uniform differentiability

Sometimes we need uniform continuity of the mapping $f' : E \to L(E,F)$ (see, for instance, Eells [1], p.760) and sometimes we need uniform differentiability of the given mapping (see, for instance, §2.1). These two notions are closely related as we shall see in this section. As usual, we assume that

E,F ∈ TLS, a ∈ A ∈ $\mathcal{O}$(E) and f : A → F.

[1.10.1] f is said to be <u>uniformly differentiable on</u> M ⊂ A if f ∈ D(M,F) and, for any V ∈ N(F) and B ∈ $\mathcal{B}$(E), there exists α > 0 such that

$$\varepsilon^{-1} r(f,x,\varepsilon z) \in V \text{ if } x \in M, z \in B \text{ and } 0 < |\varepsilon| < \alpha.$$

(1.10.2) <u>If</u> F ∈ LCS, f ∈ D(A,F) <u>and</u> f' : A → L(E,F) <u>is uniformly continuous, then</u> f <u>is uniformly differentiable on</u> A.

PROOF. For any V ∈ N(F) and B ∈ $\mathcal{B}$(E), we have, by the assumption, that, for V_1 ∈ N(F) such that $\bar{V}_1$ ⊂ V, there exists U ∈ N(E) such that

$$f'(x) - f'(y) \in (B, V_1) \text{ if } x,y \in A \text{ and } x - y \in U.$$

Take α > 0 such that αB ⊂ U. Then, if 0 < |ε| < α, z ∈ B and x ∈ A,

$$\varepsilon^{-1} r(f,x,\varepsilon z) \in |\overline{Co}| \{f'(x+\theta\varepsilon z)(z) - f'(x)(z) : 0 \leq \theta \leq 1\} \subset \bar{V}_1 \subset V.$$

The converse is more complicated.

(1.10.3) <u>Assume that</u> f ∈ D(A,F) <u>and, for some bounded absolutely convex set</u> B ⊂ A, f'(B) ∈ $\mathcal{B}$(L(E,F)). <u>If</u> f <u>is uniformly differentiable on</u> B, <u>then, for any</u> V ∈ N(F), <u>there exist</u> α,β ∈ (0,1) <u>such that</u>

$$f'(x) - f'(y) \in (B,V) \text{ if } x - y \in \alpha B \text{ and } x,y \in \beta B.$$

PROOF. By the assumption, for any V ∈ N(F), there exists ε ∈ (0,1) such that

$$\varepsilon^{-1}[f(x+\varepsilon z) - f(x)] - f'(x)(z) \in V \text{ if } x,z \in B.$$

Put β = 1 - ε. Since f'(B) ∈ $\mathcal{B}$(L(E,F)), for V_1 ∈ N(F) such that $\bar{V}_1$ ⊂ V, there is ε ∈ (0,1) such that f'(B) ⊂ (αB,εV). Then, if x - y ∈ αB and x,y ∈ B, we have

$$f(x) - f(y) \in |\overline{Co}| \{f'(y+\theta(x-y))(x-y) : 0 \leq \theta \leq 1\} \subset \varepsilon V.$$

Thus, if $x,y \in \beta B$ and $x - y \in \alpha B$, for any $z \in B$ we have

$$f'(x)(z) - f'(y)(z)$$
$$= (\varepsilon^{-1}[f(y+\varepsilon z)-f(y)] - f'(y)(z) + (\varepsilon^{-1}[f(x+\varepsilon z)-f(z)] - f'(x)(z))$$
$$+ \varepsilon^{-1}(f(x)-f(y)) - \varepsilon^{-1}(f(x+\varepsilon z)-f(y+\varepsilon z)) \in 4V,$$

or, $f'(x) - f'(y) \in (B, 4V)$.

When the spaces are normed, we have a simpler statement.

(1.10.4) <u>Let</u> $E, F \in NLS$. <u>If</u> f <u>is uniformly differentiable on</u> B_α, <u>then</u> f' <u>is uniformly continuous on</u> B_β <u>for</u> $\beta < \alpha$.

PROOF. By the assumption, for any $\varepsilon > 0$ there exists $\delta > 0$ such that $\|z\| < \delta$ implies

$$\|f(x+z) - f(x) - f'(x)(z)\| < \frac{1}{6}\varepsilon\|z\| \quad \text{if} \quad x \in B_\alpha.$$

Let $\|x-y\| < \frac{1}{2}\delta$ and $x, y \in B_\beta$. Then, for any z such that $\|z\| = 1$,

$$\|x-y+\tfrac{1}{2}\delta z\| < \delta \quad \text{and} \quad \|x+\tfrac{1}{2}\delta z\| < \alpha,$$

since we can assume $\delta < \alpha - \beta$. Therefore,

$$\|f'(x) - f'(y)\| \leq \frac{2}{\delta} \sup_{\|z\|=1} \{\|f'(x)(\tfrac{1}{2}\delta z) + f(x) - f(x+\tfrac{1}{2}\delta z)\|$$
$$+ \|f(x+\tfrac{1}{2}\delta z) - f(y) - f'(y)(x-y+\tfrac{1}{2}\delta z)\|$$
$$+ \|f(x) - f(y) - f'(y)(x-y)\|\} < \varepsilon.$$

NOTE: In the case of NLS, these facts have been discussed in Vainberg [1], Chapter 1, §4. See also Dieudonné [1], Chapter VIII, §6, problem 3b).

§1.11 Partial derivatives

Let $E_1, E_2 \in TLS$, $E = E_1 \times E_2$, $a = (a_1, a_2) \in A \in \mathcal{O}(E)$ and $f: A \to F$. We consider families M_i on E_i ($i=1,2$) as in §1.2 and put

$$M = \{M_1 \times M_2 : M_i \in M_i \ (i=1,2)\} .$$

[1.11.1] f is said to be <u>partially</u> M-<u>differentiable at</u> a <u>in the first</u> <u>variable</u> if the mapping $x_1 \mapsto f(x_1,x_2)$ of $E \times \{a_2\}$ into F is M_1-differentiable at a_1 and, if this is the case, the derivative is denoted by $\partial_1 f(a_1,a_2)$. The partial derivative $\partial_2 f(a_1,a_2)$ of f at a in the second variable is defined similarly.

By the definition, $\partial_i f(a_1,a_2) \in L(E_i,F)$ (i=1,2).

(1.11.2) <u>If</u> $f \in D_M(a,F)$, <u>then</u> $\partial_i f(a_1,a_2)$ (i=1,2) <u>exist and</u>

$$f'(a_1,a_2)(x_1,x_2) = \partial_1 f(a_1,a_2)(x_1) + \partial_2 f(a_1,a_2)(x_2) .$$

PROOF. Since $f'(a_1,a_2) \in L(E_1 \times E_2, F)$, there are $u_i \in L(E_i,F)$ (i=1,2) such that $f'(a_1,a_2)(x_1,x_2) = u_1(x_1) + u_2(x_2)$ for all $(x_1,x_2) \in E_1 \times E_2$. Let $M_1 \in M_1$ and put $M = M_1 \times \{0\}$. From the assumption, for any $V \in N(F)$ there is $\alpha > 0$ such that, if $0 < |\epsilon| < \alpha$,

$$\epsilon^{-1}[f(a_1+\epsilon x_1, a_2+\epsilon x_2) - f(a_1,a_2)] - f'(a_1,a_2)(x_1,x_2) \in V$$

for any $(x_1,x_2) \in M$, which is equivalent to

$$\epsilon^{-1}[f(a_1+\epsilon x_1, a_2) - f(a_1,a_2)] - u_1(x_1) \in V \quad \text{if} \quad x_1 \in M_1 .$$

Thus, $\partial_1 f(a_1,a_2)$ exists and equals u_1. Similarly, $\partial_2 f(a_1,a_2)$ exists and equals u_2.

The converse of this theorem is, as is well-known, not true even in two-dimensional spaces. However, if we assume continuous differentiability, we have a simpler relation.

(1.11.3) <u>Let</u> $F \in$ LCS. <u>Then</u>, $f \in C^1(A,F)$ <u>if and only if</u> $\partial_i f : A \to L(E_i,F)$ (i=1,2) <u>exist and continuous on</u> A.

PROOF. ⇒. By the above theorem, $\partial_i f$ (i=1,2) exist. To prove the continuity of $\partial_1 f$, take $(B,V) \in N(L(E,F))$. Then, by the assumption, there exists $U = U_1 \times U_2 \in N(E)$ such that

$$f'(a_1+x_1, a_2+x_2) - f'(a_1,a_2) \in (B,V) \quad \text{if} \quad (x_1,x_2) \in U.$$

Thus, for $B = B_1 \times \{0\}$ for some $B_1 \in B(E_1)$, this implies

$$\partial_1 f(a_1+x_1, a_2+x_2) - \partial_1 f(a_1,a_2) \in (B,V) \quad \text{if} \quad (x_1,x_2) \in U,$$

which implies the continuity of $\partial_1 f$. Similarly, $\partial_2 f$ is continuous.

⇐. To prove that $f \in D(a,F)$, we take $V \in N(F)$ and $V_1 \in N(F)$ such that $\bar{V}_1 + \bar{V}_1 \subset V$. Let $B = B_1 \times B_2 \in B(E)$. Then, there exists $U \in N(E)$ such that $(x_1,x_2) \in U$ implies

$$\partial_i f(a_1+x_1, a_2+x_2) - \partial_i f(a_1,a_2) \in (B_i, V_1) \quad (i=1,2).$$

Then, for any $(x_1,x_2) \in B$, since $(\varepsilon_1 x_1, \varepsilon_2 x_2) \in U$ if $|\varepsilon_1|$ and $|\varepsilon_2|$ are small, we have

$$\varepsilon^{-1}[f(a_1+\varepsilon x_1, a_2+\varepsilon x_2) - f(a_1,a_2)] - \partial_1 f(a_1,a_2)(x_1) - \partial_2 f(a_1,a_2)(x_2)$$

$$= \varepsilon^{-1}[f(a_1+\varepsilon x_1, a_2+\varepsilon x_2) - f(a_1, a_2+\varepsilon x_2)] - \partial_1 f(a_1,a_2)(x_1)$$

$$+ \varepsilon^{-1}[f(a_1, a_2+\varepsilon x_2) - f(a_1,a_2)] - \partial_2 f(a_1,a_2)(x_2)$$

$$\in |\overline{Co}|\{\partial_1 f(a_1+\theta_1 \varepsilon x_1, a_2+\varepsilon x_2)(x_1) - \partial_1 f(a_1,a_2)(x_1) : 0 \leq \theta_1 \leq 1\}$$

$$+ |\overline{Co}|\{\partial_2 f(a_1, a_2+\theta_2 \varepsilon x_2)(x_2) - \partial_2 f(a_1,a_2)(x_2) : 0 \leq \theta_2 \leq 1\}$$

$$\subset \bar{V}_1 + \bar{V}_1 \subset V.$$

Thus, $f \in D(A,F)$. The continuity of $f' : A \to L(E,F)$ is obvious.

(1.11.4) <u>Let</u> $F \in LCS$. <u>Let</u> f <u>be partially Gâteaux differentiable in the first (or second) variable in</u> A. <u>Then, for</u> $\xi, \eta \in R$ <u>and</u> $x_i \in E_i$ $(i=1,2)$ <u>such that</u> $(a_1+\xi x_1, a_2+\eta x_2) \in A$ <u>and for</u> $\bar{a} \in \bar{F}$, <u>the function</u>

$$\phi(\xi,\eta) = \langle f(a_1+\xi x_1, a_2+\eta x_2), \bar{a} \rangle$$

is partially differentiable in the first (or second) variable and

$$\frac{\partial \phi}{\partial \xi} = < \partial_1 f(a_1+\xi x_1, a_2+\eta x_2)(x_1), \bar{a} >$$

$$(\text{or} \quad \frac{\partial \phi}{\partial \eta} = < \partial_2 f(a_1+\xi x_1, a_2+\eta x_2)(x_2), \bar{a} >) .$$

PROOF.   Straightforward and omitted.

We shall next present two sufficient conditions for the commutativity of higher partial derivatives.   The proofs are immediate consequences of the two-dimensional cases via (1.11.4).

(1.11.5)   Let $F \in LCS$. Assume that $\partial_i f(x_1, x_2)$ (i=1,2) exist for all $(x_1, x_2) \in A$.

1°.   If $\partial_i f \in D_H((a_1, a_2), L(E_i, F))$ (i=1,2), then $\partial_1 \partial_2 f(a_1, a_2)$ and $\partial_2 \partial_1 f(a_1, a_2)$ exist.

2°.   If $\partial_1 \partial_2 f(a_1, a_2)$ exists in a neighbourhood of (a_1, a_2) and continuous there, then $\partial_2 \partial_1 f(a_1, a_2)$ exists.

Moreover, in both cases, we have $\partial_1 \partial_2 f(a_1, a_2) = \partial_2 \partial_1 f(a_1, a_2)$.

NOTE: As is well-known, (1.11.5). 1° is called Young's theorem (Young [1]) and 2° is called Schwartz's theorem.   1° is sometimes more convenient than 2°, because it does not need the continuity which, as we have seen in §1.7, is one of the difficulties in the calculus in TLS.

§1.12 Other differentiations

During its history of more than thirty years, differentiability in TLS has been given various definitions.   The aim of this section is to exhibit these definitions and determine the relations among them.   All the material has been taken from Averbukh-Smolyanov [2].   We shall omit the proofs.

Throughout, we assume that $E, F \in LCS$, although it is possible to give equivalent definitions in general TLS.   To avoid repetition, we shall always assume that f is Gâteaux differentiable and we shall denote the remainder

$r(f,a,x)$ of f at a with respect to the Gateaux derivative $f'(a)$ by $r(x)$. As usual, $a \in A \in \mathcal{O}(E)$ and $f : A \to F$.

There are five conditions

I. For any $q \in P(F)$ there exists $p \in P(E)$ such that the following condition is satisfied : for any $\varepsilon > 0$ there exists $\delta > 0$ such that $p(x) \leq \delta$ implies $q(r(x)) \leq \varepsilon p(x)$.

This definition is due to Hyers [1], and it is equivalent to the following one, which was adopted by Lang [1] : "For any $V \in N(F)$ there exists $U \in N(E)$ such that $x \in U$ implies $r(\varepsilon x) \in o(\varepsilon)V$".

II. There exists $p \in P(E)$ such that the following condition is satisfied : for any $q \in P(F)$ and any $\varepsilon > 0$ there exists $U \in N(E)$ such that $x \in U$ implies $q(r(x)) \leq \varepsilon p(x)$.

This is due to Keller [1] and is equivalent to the one given by Michal [2] which is as follows : "There exists a mapping $\mu : E \times E \to F$ such that $1°$. $\mu(x,x) = r(x)$ for any $x \in E$; $2°$. $\mu(0,x) = 0$ for any $x \in E$; $3°$. $\mu(x_1, \xi x_2) = \xi \mu(x_1, x_2)$ if $\xi \in R$ and $x_1, x_2 \in E$, and $4°$. there exists U_0 such that the following condition is satisfied : for any $V \in N(F)$ there is $U \in N(E)$ such that $x_1 \in U_1$ and $x_2 \in U_0$ imply $\mu(x_1, x_2) \in V$".

III. For any $q \in P(F)$ there exists $p \in P(E)$ such that the following condition is satisfied : For any $\varepsilon > 0$ there exists $U \in N(E)$ such that $x \in U$ implies $q(r(x)) \leq \varepsilon p(x)$.

This is due to Marinescu [2] and Sebastião e Silva [1]. Keller [1] also gave the same definition.

IV. For any $V \in N(F)$ and any $x_0 \in E$ there exist $U \in N(E)$ and $\delta > 0$ such that $x \in x_0 + U$, $0 < |\varepsilon| < \delta$ imply $\varepsilon^{-1} r(\varepsilon x) \in V$.

This is due to Bastiani [1] and is equivalent to the following one given by Michal [1] : "There exists a mapping $\mu : E \times E \to F$ which satisfies $1°, 2°, 3°$ of II and $4°$. μ is continuous on $\{0\} \times E$".

V. For any $B \in \mathcal{B}(E)$ there exists $C \in \mathcal{B}(F)$ such that the following condition is satisfied : for any $\alpha > 0$ there exists $\delta > 0$ such that $x \in B$, $0 < |\varepsilon| < \delta$ imply $\varepsilon^{-1} r(\varepsilon x) \in C$.

The following diagram is due to Averbukh-Smolyanov [2].

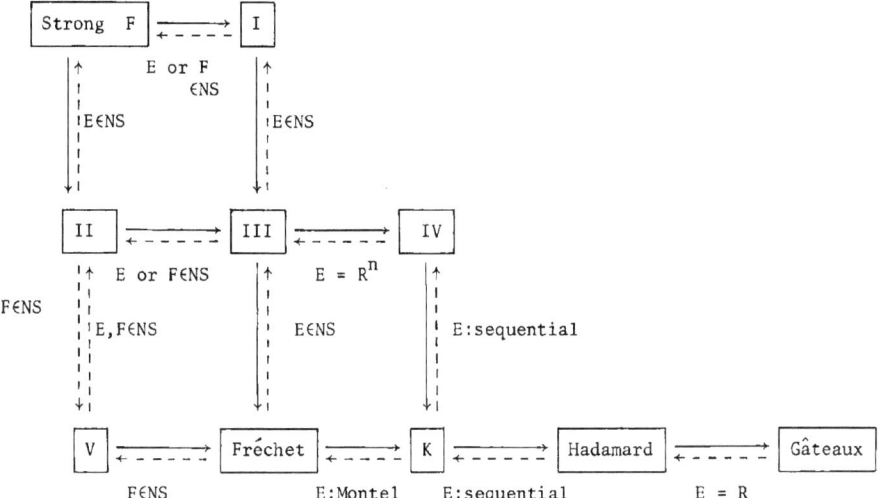

Guide : 1°. ⟶ means the left-hand differentiability implies right-hand one.

2°. ←-- means the same under the conditions written there.

3°. K is the M-differentiability with $M = K$.

About this diagram, we note the following :

1°. The differentiabilities in the first and second rows imply continuity. Others do not.

2°. Among the differentiabilities that coincide with the Fréchet differentiability when $E, F \in$ NLS, the strongest one is our strong Fréchet differentiability and the weakest is our Fréchet differentiability.

Negative examples.

1. The mapping $f : R^{\infty} \to R^{\infty}$ defined by

$$f(x_1, x_2, \ldots) = (x_1^2, x_2^2, \ldots)$$

is differentiable at 0 in the sense of I, but not strongly Fréchet differentiable. This example is due to Keller [1].

2. The mapping $f : R^\infty \to R$ defined by

$$f(x_1, x_2, \ldots) = x_1 \sum_{i=1}^{\infty} 2^{-i} |x_i|/(1+|x_i|)$$

is everywhere differentiable in the sense of II, but nowhere so in the sense of I.

3. Let $E \in BS$ and regard it as a LCS with the weak topology. Then, the mapping $f : E \to R$ defined by $f(x) = \|x\|^2$ is everywhere Fréchet differentiable, but nowhere differentiable in the sense of III.

4. Let $C(R)$ be the inductive limit of $C(-n,n)$. For $x, y \in C(R)$, put

$$(x,y) = \int_{-\infty}^{\infty} x(t)y(t)dt \quad \text{and} \quad \|x\|^2 = (x,x) .$$

Let $e_i \in C(R)$ be such that $\|e_i\| = 1$, $(e_i, e_j) = 0$ if $i \neq j$, and $e_i(t) = 0$ if $|t| \geq n$ for some n. Let $a \in C(R)$ be such that $\text{supp}(a(t)) \subset [n, n+1]$. Further, for

$$x_{m,k}(t) = \frac{1}{m} e_k(t) + \frac{1}{k} a(t-m) \quad (k,m=1,2,\ldots)$$

and, for any $x \in C(R)$, put

$$f_{m,k}(x) = \begin{cases} \frac{1}{m} \exp(-1/(\|x-x_{m,k}\|^2 - \frac{1}{3m^2})) & \text{if } \|x\| < \|x_{m,k}\| ; \\ 0 & \text{otherwise.} \end{cases}$$

Then, the mapping

$$f(x) = \sum_{m,k} f_{m,k}(x)$$

is continuous and C^∞ in the sense of Fréchet, but is not so at 0 in the sense of IV.

5. The mapping $f : \ell^\infty \to R^\infty$ defined by $f((x_i)) = (y_i)$, where

$$y_i = x_i | (1-ix_i) \quad \text{if } 1 - ix_i \neq 0; = 0 \text{ otherwise,}$$

is Fréchet differentiable at 0, but is not so in the sense of V.

CHAPTER 2. COMPACT MAPPINGS

The aim of this chapter is to study the relations between the compactness of a mapping and the compactness of its derivatives. The fact that the Fréchet derivative of a compact mapping is also compact has played an important rôle in, for example, the variational theory (see Vainberg [1]) and the Leray-Schauder theory of mapping degree (see Krasnoseliski [1]).

Throughout, we assume that

$$E, F \in TLS, \ a \in A \in \mathcal{O}(E) \ \text{ and } \ f : A \to F.$$

§2.1 Compact mappings and their Fréchet derivatives

We define the compactness of a mapping as follows.

[2.1.1] f is said to be $\mathcal{B}$-compact ($\mathcal{B}$-precompact) on A if, for any $B \in \mathcal{B}(E)$ such that $B \subset A$, $f(B)$ is contained in a compact (precompact) subset of F.

f is said to be compact at a (precompact at a) if there exists $U_a \in \mathcal{N}(E)$ such that $a + U_a \subset A$ and $f(a+U_a)$ is contained in a compact (precompact) subset of F. If f is compact at every point of A, f is simply said to be compact on A.

It is obvious that, when $f : E \to F$ is linear, then f is compact (precompact) if and only if it is so at 0. Moreover, every compact linear mapping is continuous. However, not every $\mathcal{B}$-compact linear mapping is continuous, as we can easily see if we put $F = R$.

For non-linear mappings, compactness does not imply $\mathcal{B}$-compactness. In fact, the function $f : R \setminus \{0\} \to R$ defined by $f(\xi) = \xi^{-1}$ is compact but not $\mathcal{B}$-compact. Moreover, contrary to the linear case, a compact non-linear mapping is not necessarily continuous even when $E = F = R$, as every step function is compact.

In the following, we shall mainly be concerned with pre compactness. The corresponding statements for compactness can be obtained if we add completeness to the assumptions.

We shall begin with $\mathcal{B}$-precompactness.

(2.1.2) Let E,F $\in$ LCS. For f $\in$ D(E,F), we consider the following properties :

(1) f is $\mathcal{B}$-precompact;

(2) f'(x) is $\mathcal{B}$-precompact for every x $\in$ E;

(3) for any $B_1, B_2 \in \mathcal{B}(E)$, $f'(B_1)(B_2)$ is precompact;

(4) the mapping f' : E $\to$ L(E,F) is $\mathcal{B}$-precompact.

Then, we have

1°. (2) + (4) $\Rightarrow$ (3) $\Rightarrow$ (1) $\Rightarrow$ (2).

2°. If f is uniformly differentiable on every $B \in \mathcal{B}(E)$, then (1) $\Rightarrow$ (3).

3°. (2) does not imply (1) even when E and F are Hilbert spaces.

PROOF. 1°. (1) $\Rightarrow$ (2).

Let a $\in$ E and put

$$f_\varepsilon(x) = \varepsilon^{-1}[f(a+\varepsilon x) - f(a)] \quad \text{for} \quad \varepsilon \neq 0 \text{ and } x \in E.$$

Then, the mapping f_ε : E $\to$ F is $\mathcal{B}$-precompact and converges to f'(a) uniformly on each bounded set. Let $B \in \mathcal{B}(E)$, $V \in N(F)$, $V_1 \in N(F)$ and $V_1 + V_1 \subset V$. Then, there exists ε such that $(f_\varepsilon - f'(a))(B) \subset V_1$. Thus, if
$f_\varepsilon(B) \subset \bigcup_{i=1}^n (f_\varepsilon(x_i) + V_1)$, by precompactness of $f_\varepsilon(B)$, we have
$f'(a)(B) \subset f_\varepsilon(B) + V_1 \subset \bigcup_{i=1}^n (f_\varepsilon(x_i) + U)$.

(3) $\Rightarrow$ (1).

For any $B \in \mathcal{B}(E)$ and x $\in$ B,

$$f(x) - f(0) \in |\overline{Co}|\{f'(\theta x)(x) : 0 \leq \theta \leq 1\} \subset |\overline{Co}|\{f'(B_1)(B)\},$$

where B_1 is the circled cover of B. Since $f'(B_1)(B)$ is precompact, its closed convex cover is also precompact. Hence, f(B) is precompact.

(2) + (4) ⇒ (3).

Let $V, V_1 \in N(E)$ and $V_1 + V_1 \subset V$. Since $f'(B_1)$ is precompact in $L(E,F)$, there are $x_i \in B_1$ (i=1,2,...,n) such that $f'(B_1) \subset \bigcup_{i=1}^{n} \{f'(x_i) + (B_2, V_1)\}$. Since, for each i, $f'(x_i)(B_2)$ is precompact, there are $y_{i,j} \in B_2$ (j=1,2,...,m) such that $f'(x_i)(B_2) \subset \bigcup_{j=1}^{m} \{f'(x_i)(y_{i,j}) + V_1\}$. Thus,

$$f'(B_1)(B_2) \subset \bigcup_{i,j=1}^{m,n} \{f'(x_i)(y_{i,j}) + V\}.$$

2°. For $B_1, B_2 \in B(E)$ and $V \in N(F)$, there exists $\delta > 0$ such that

$$\delta^{-1}[f(x+\delta z) - f(x) - f'(x)(\delta z)] \in V \text{ if } x \in B_1 \text{ and } z \in B_2.$$

Therefore, $f'(B_1)(B_2) \subset \delta^{-1}[f(B_1 + \delta B_2) - f(B_1)] + V$. Since the first member of the right-hand side is precompact, $f'(B_1)(B_2)$ is also precompact.

3°. In ℓ^2, consider the mapping $f : (\xi_n) \mapsto (\xi_n^2)$. For the coordinate elements e_n, we have $f(e_n) = e_n$, which means that f is not B-precompact. For $a = (\alpha_n) \in \ell^2$, consider the mapping $u : (\xi_n) \mapsto (2\alpha_n \xi_n)$. Obviously, u is linear and

$$\|\|x\|^{-1}[f(a+x) - f(a) - u(x)]\| \leq \|x\|,$$

which means that f is Fréchet differentiable and $f'(a) = u$. It is easy to see that $f'(a)$ is B-precompact.

If we take neighbourhoods instead of bounded sets, we have strong Fréchet differentiation and compactness at a point defined above. Between these two notions, we have the same relations as those in (2.1.2).

There is another notion which is closely related to the B-compactness considered above.

[2.1.3] Let $E, F \in LCS$. $f : E \to F$ is said to be **strongly continuous** if it is continuous as a mapping of E with its weak topology into F with its original topology.

Obviously, if E is semi-reflexive and f is strongly continuous, then f is β-compact. However, the converse is not true.

(2.1.4) Let E,F ∈ LCS and f ∈ D(E,F). We consider the following properties :
   (1)   f is strongly continuous;
   (2)   f'(x) is strongly continuous for each x ∈ E;
   (3)   f'(x)(y) is jointly strongly continuous;
   (4)   f' : E → L(E,F) is strongly continuous.

Then, on any bounded set, we have the following relations :
   1°.   (2) + (4) ⇒ (3) ⇒ (1) ⇒ (2).
   2°.   If f is uniformly differentiable on any bounded set, then (1) ⇒ (3).
   3°.   If E is semi-reflexive, (3) ⇒ (4).
   4°.   (2) does not imply (1).

PROOF is similar to that of (2.1.2). However, we shall add some remarks.

(1) To see that 2° fails without some restrictions such as the uniform differentiability, one may consider the mapping f : R → R defined by $f(\xi) = \xi^2 \sin(1/\xi)$ if $\xi \neq 0$; $= 0$ if $\xi = 0$.

(2) The same example as in (2.1.2). 4° can be used for 4° here.

(3) The following fact is a corollary : Let E be semi-reflexive and f : E → F be uniformly differentiable on any bounded set. Then, f is strongly continuous if and only if f'(x) : E → F for each x ∈ E and f' : E → L(E,F) are all β-precompact. This fact has been proved by Palmer [1].

NOTE: The proof of (1) ⇒ (2) in (2.1.2) can be found in Krasnoseliski [1] and Vainberg [1] in the case of NLS and in J.T. Schwartz [1] in the case of TLS. The fact (2) + (4) ⇒ (1) in the case of NLS is due to Vainberg [1]. The example 3° has been given by Banic [1] and Yamamuro [2]. Other results are due to Lloyd [1]. Results related to the strong continuity are essentially due to Palmer [1]. For the corresponding results for families of mappings, see Moore [1] and Lloyd [1].

§2.2 Compact mappings and Hadamard differentiability

We shall, at first, show that the Hadamard derivative of a B-compact mapping is not necessarily B-compact. To do this, we consider the space ℓ^2. Recall that a subset K is compact if and only if

  1°. K is bounded;

  2°. $\lim\limits_{n\to\infty} \sup\limits_{x\in K} \sum\limits_{i=n}^{\infty} \xi_i^2 = 0$ where $x = (\xi_i)$.

Now, consider the set

$$K_0 = \{x \in \ell^2 : |\xi_n| \leq 1/n \text{ for every } n\},$$

and the mapping $f : (\xi_n) \mapsto (\phi_n(\xi_n))$, where ϕ_n are differentiable functions of R into R such that

$$\phi_n(\xi) = \xi \text{ if } |\xi| \leq 1/2n; = 0 \text{ if } |\xi| > 1/n.$$

Since $|\phi_n(\xi)| \leq 1/n$, f maps ℓ^2 into K_0, and, hence, it is B-compact and compact everywhere. Moreover, it is continuous. In fact, for any $\varepsilon > 0$, take n_0 such that $\sum\limits_{n=n_0+1}^{\infty} (\frac{2}{n})^2 < \varepsilon$. Then, if $x_i \to x_0$, we have

$$\|f(x_i) - f(x_0)\|^2 = \sum_{n=1}^{\infty} (\phi_n(\xi_n^{(i)}) - \phi_n(\xi_n^{(0)}))^2 \quad \text{where } x_i = (\xi_n^{(i)})$$

$$\leq \sum_{i=1}^{n_0} (\phi_n(\xi_n^{(i)}) - \phi_n(\xi_n^{(0)}))^2 + \varepsilon.$$

Further, $f \in D_H(0, \ell^2)$ and $f'(0) = 1$. To show this, let K be any compact set. For any $\delta > 0$, take n_0 such that

$$\sup_{x\in K} \sum_{n=n_0+1}^{\infty} \xi_n^2 < \delta \quad \text{where } x = (\xi_n).$$

Since $\|x\| \leq \gamma$ for some γ whenever $x \in K$, $|\xi_n| \leq \gamma$ for all $(\xi_n) \in K$. Therefore, $\sup\limits_{x\in K} \max(|\xi_1|, \ldots, |\xi_{n_0}|) \leq \gamma$. Hence, there exists $\varepsilon_0 > 0$ such that $|\varepsilon| \leq \varepsilon_0$ implies

$$|\varepsilon| \max(|\xi_1|,\ldots,|\xi_{n_0}|) \leq 1/2n_0 \quad \text{for all} \quad x \in K,$$

which means that

$$|\varepsilon\xi_n| \leq 1/2n_0 \leq 1/2n \quad \text{if} \quad n = 1,2,\ldots,n_0 ,$$

which is equivalent to $\phi_n(\xi_n) = \varepsilon\xi_n$ if $n = 1,2,\ldots,n_0$. Therefore,

$$\|\varepsilon^{-1}f(\varepsilon x) - x\|^2 = \left(\sum_{n=1}^{n_0} + \sum_{n=n_0+1}^{\infty}\right)(\varepsilon^{-1}\phi_n(\varepsilon\xi_n)-\xi_n)^2$$

$$= \sum_{n=n_0+1}^{\infty}(\varepsilon^{-1}\phi_n(\varepsilon\xi_n)-\xi_n)^2 \leq \sum_{n=n_0+1}^{\infty}\xi_n^2 < \delta ,$$

if $x \in K$ and $0 < |\varepsilon| < \varepsilon_0$.

There is another phenomenon which relates Hadamard differentiation to compact mappings. The following theorem shows, by making use of compact mappings, that Hadamard differentiation is, in a sense, the weakest differentiation which satisfies the composition property.

(2.2.1) <u>Let</u> $E \in LCS$ <u>and</u> $f : E \to E$.

1°. <u>If</u> $f \in D_H(E,E)$, <u>then</u> $f \circ k \in D(E,E)$ <u>for any</u> B-<u>compact</u> $k \in D(E,E)$ and $(f \circ k)'(x) = f'(k(x)) \circ k'(x)$ <u>for all</u> $x \in E$.

2°. <u>If</u> $f \in D_G(E,E)$ <u>and if, for any</u> B-<u>compact</u> $k \in D(E,E)$, <u>we have</u> $f \circ k \in D_G(E,E)$ <u>and the composition formula holds, then</u> $f \in D_H(E,E)$.

PROOF. 1°. By (1.2.9), $f \circ k \in D_H(E,E)$ and the composition formula holds. It is easy to see that

$$r(f \circ k, a, x) = f'(k(a))(r(k,a,x)) + r(f,k(a),k(a+x)-k(a)) .$$

Therefore, what we have to show is that

$$\varepsilon^{-1}r(f,k(a),k(a+\varepsilon x)-k(a)) \to 0$$

uniformly on any bounded set. Assume that this is not true. Then there exist $\{\varepsilon_n\} \in c_0$ and $\{x_n\} \in B(E)$ such that

$$\varepsilon_n^{-1} r(f,k(a)), \varepsilon_n \varepsilon_n^{-1}(k(a+\varepsilon_n x_n) - k(a))) \not\to 0.$$

However, since

$$\varepsilon_n^{-1}(k(a+\varepsilon_n x_n)-k(a)) = k'(a)(x_n) + \varepsilon_n^{-1} r(k,a,\varepsilon_n x_n),$$

and, by (2.1.2). 1°, the set $\{k'(a)(x_n)\}$ is relatively compact, the set $\{\varepsilon_n^{-1}(k(a+\varepsilon_n x_n) - k(a))\}$ is relatively compact. This contradicts the fact that $f \in D_H(a,E)$.

2°. Assume that $f \notin D_H(a,E)$. Then there exist $\{\varepsilon_n\} \in c_0$ and $\{x_n\} \subset E$ such that $x_n \to x_0$ and $\varepsilon_n^{-1} r(f,a,\varepsilon_n x_n) \not\to 0$. Now, by the same method as in (1.2.7), we have an E-valued function $s(\xi)$ of a real variable such that

$$s(0) = a, \ s(\varepsilon_n) = a + \varepsilon_n x_n \text{ and } s'(0) = x_0.$$

Then consider the mapping $k : E \to E$ defined by

$$k(x) = s(<x,\tilde{a}>),$$

where $\tilde{a} \in \tilde{E}$ such that $<a,\tilde{a}> = 1$. By the assumption, $f \circ k \in D_G(0,E)$ and $(f \circ k)'(0) = f'(k(0)) \circ k'(0)$. On the other hand,

$$\varepsilon_n^{-1} r(f,a,\varepsilon_n x_n) = \varepsilon_n^{-1}[f(a+\varepsilon_n x_n) - f(a) - f'(a)(\varepsilon_n x_n)]$$

$$= \varepsilon_n^{-1}[(f \circ k)'(\varepsilon_n a) - (f \circ k)(0) - (f \circ k)'(0)(\varepsilon_n a)] + (f \circ k)'(0)(a) - f'(a)(x_n)$$

$$= \varepsilon_n^{-1} r(f \circ k, 0, \varepsilon_n a) + f'(a)(x_0 - x_n) \to 0,$$

which is a contradiction.

NOTE: The fact that Hadamard differentiation is the weakest differentiation which has the composition property has first been proved by Averbukh-Smolyanov [2] by a more direct way. (2.2.1) was proved in Yamamuro [3].

CHAPTER 3. INVERSE MAPPING THEOREMS

The inverse mapping theorems are the most important tools which connect the differential calculus with the investigation of topological properties. In the case of Banach space, as we shall see, we have satisfactory and useful results. However, if the space is a general TLS, or even a non-normed MLS, the straightforward generalization of Banach space results is no longer possible.

§3.1 Differentiation in $L(E,F)$

Let $E, F \in$ TLS. We recall that $L(E,F)$, the set of all continuous linear mappings of E into F, is equipped with the topology of uniform convergence on bounded sets.

In this section, we shall consider the differentiability of the following two mappings :

composition operation :   comp : $L(E,F) \times L(F,G) \to L(E,G)$ defined by

$$\text{comp}(u,v) = v \circ u \quad \text{for} \quad u \in L(E,F) \quad \text{and} \quad v \in L(F,G),$$

and

inverse operation : inv : $\text{Isom}(E,F) \to \text{Isom}(F,E)$ defined by

$$\text{inv}(u) = u^{-1} \quad \text{for} \quad u \in \text{Isom}(E,F).$$

As can be seen from Appendix 2, the mapping comp is not always continuous. Therefore, it is not differentiable by any differentiation that implies the continuity, such as the strong Fréchet differentiation defined in [1.7.3]. However, we have the following result.

(3.1.1) The mapping comp is Fréchet differentiable everywhere and, for (u,v), $(u_0, v_0) \in L(E,F) \times L(F,G)$, we have

$$(\text{comp})'(u_0, v_0)(u,v) = v_0 u + v u_0.$$

PROOF. Since comp is separately continuous, the mapping $(u,v) \to v_0 u + v u_0$ of $L(E,F) \times L(F,G)$ into $L(E,G)$ is continuous, and it is evidently linear. We shall show that this is the Fréchet derivative of comp at (u_0, v_0). For bounded sets $\tilde{B}_1 \subset L(E,F)$ and $\tilde{B}_2 \subset L(F,G)$, the set $\tilde{B}_2 \circ \tilde{B}_1$ is bounded in $L(E,G)$. Hence,

$$\varepsilon^{-1} r(\text{comp}, (u_0, v_0), \varepsilon(u,v)) = \varepsilon v u$$

converges to zero uniformly for $(u,v) \in \tilde{B}_1 \times \tilde{B}_2$.

The following fact was found by John Lloyd.

(3.1.2) *The mapping comp is* C^1 *if every bounded subset of* $L(F,G)$ *is equicontinuous*.

PROOF. We need to show that the mapping

$$(\text{comp})' : L_1 \times L_2 \to L(L_1 \times L_2, L_3)$$

is continuous at $(0,0)$, where

$$L_1 = L(E,F), \quad L_2 = L(F,G) \quad \text{and} \quad L_3 = L(E,G) .$$

We take an arbitrary $\hat{W} \in N(L(L_1 \times L_2, L_3))$. We can assume that

$$\hat{W} = (\hat{B}_1 \times \hat{B}_2, (B, W)) ,$$

where $\tilde{B}_i \in B(L_i)$ (i=1,2), $B \in B(E)$ and $W \in N(G)$. Let $W_1 \in N(G)$ be such that $W_1 + W_1 \subset W$, and put

$$B_1 = \tilde{B}_1(B) .$$

Then, $B_1 \in B(F)$ and, hence, $(B_1, W_1) \in N(L_2)$. Moreover,

$$v(\tilde{B}_1(B)) \subset W_1 \quad \text{if} \quad v \in (B_1, W_1) .$$

By the assumption, $\tilde{B}_2$ is equicontinuous. Hence, we can find $V \in N(F)$ such that $\tilde{B}_2(V) \subset W_2$, and

$$\hat{B}_2(u(B)) \subset W_2 \quad \text{if} \quad u \in (B,V) .$$

Therefore, for $\hat{V} = (B,V) \times (B_1,W_1) \in N(L_1 \times L_2)$,

$$v(\hat{B}_1(B)) + \hat{B}_2(u(B)) \subset W \quad \text{if} \quad (u,v) \in \hat{V} ,$$

which means that

$$(\text{comp})'(u,v) \in \hat{W} \quad \text{if} \quad (u,v) \in \hat{V}.$$

If F is <u>quasi-barrelled</u>, every bounded subset of L(F,G) is equicontinuous. Conversely, if every bounded subset of $L(F,R) = \bar{F}$ is equicontinuous, F is quasi-barrelled.

(3.1.3) <u>If</u> F <u>is quasi-barrelled, the mapping</u> comp <u>is</u> C^∞ <u>everywhere and</u>

$$(\text{comp})''(u_0,v_0)(u,v) = (\text{comp})'(u,v) ,$$

$$(\text{comp})^{(n)}(u_0,v_0) = 0 \quad \text{for} \quad n \geq 3 .$$

PROOF. The mapping

$$(\text{comp})' : L_1 \times L_2 \to L(L_1 \times L_2, L_3)$$

is linear and it is continuous by (3.1.2). Therefore, it is differentiable and we have the first equality, which means that the mapping (comp)'' is constant. Hence, it is C^∞ and the derivatives are zeros.

Thus, for a quasi-barrelled F, the mapping comp is C^∞ everywhere, and nowhere continuous if F is not normed.

For the mapping inv, we have to restrict ourselves to the case of BS. As is well-known (e.g., Dieudonné [1], p.148), if $E \in BS$ and $w \in L(E,E)$ satisfies $\|w\| < 1$, then $1 + w \in \text{Isom}(E,E)$ and

$$(1+w)^{-1} = \sum_{n=0}^{\infty} (-1)^n w^n ,$$

where the series is absolutely convergent. It is an immediate consequence of this fact that

$$\|(1+w)^{-1} - 1 + w\| \leq \|w\|^2/(1-\|w\|) .$$

(3.1.4) <u>Let</u> $E, F \in BS$. <u>Then, the mapping</u> inv : $\mathrm{Isom}(E,F) \to \mathrm{Isom}(F,E)$ <u>is differentiable everywhere in every order and</u>

$$(\mathrm{inv})'(u_0)(u) = u_0^{-1} \circ u \circ u_0^{-1} .$$

PROOF. See Dieudonné [1], p.148 and p.179.

§3.2 Differentiability of inverse mappings

In addition to the facts that "differentiability does not imply continuity" and "comp is not always continuous", we here see the third difference between the normed and non-normed cases. To obtain a TLS version of the normed case theorem (3.2.5), we need the following definition. Throughout this section, we assume that

$$E, F \in \mathrm{TLS}, \ a \in A \in \mathcal{O}(E) \ \text{ and } \ f : A \to F .$$

[3.2.1] Let M_1 and M_2 be M-classes on E and F respectively. Then, f is said to be (M_1, M_2)-preserving at a if, for any $M_1 \in M_1$, $\{\varepsilon_n\} \in (c_0)$ and $\{x_n\} \subset M_1$, there exists $M_2 \in M_2$ such that $\varepsilon_n^{-1}[f(a+\varepsilon_n x_n) - f(a)] \in M_2$ for almost all n.

It is evident that, if f is linear, it is (M_1, M_2)-preserving at a if and only if it is so at zero, and it is equivalent to the following condition : for any $M_1 \in M_1$ there exists $M_2 \in M_2$ such that $\{x_n\} \subset M_1 \in M_1$ implies $\{f(x_n)\} \subset M_2$.

(3.2.2) <u>Assume that</u> M_1 <u>and</u> M_2 <u>are M-classes on</u> E <u>and</u> F <u>respectively</u>, $M_2 + K_s(F) \subset M_2$ <u>and</u> $f \in D_{M_1}(a,F)$. <u>Then</u>, f <u>is</u> (M_1, M_2)-<u>preserving at</u> a <u>if and only if</u> $f'(a)$ <u>is</u> (M_1, M_2)-<u>preserving</u>.

PROOF. Assume that f is (M_1,M_2)-preserving, $M_1 \in \mathcal{M}_1$ and $\{x_n\} \subset M_1$. Let $\{\varepsilon_n\} \in (c_0)$. Then, since $f \in D_{M_1}(a,F)$, we have

$$f'(a)(x_n) + \varepsilon_n^{-1} r(f,a,\varepsilon_n x_n) \in M_2$$

for some $M_2 \in \mathcal{M}_2$, and $\varepsilon_n^{-1} r(f,a,\varepsilon_n x_n) \to 0$. Put $K = \{\varepsilon_n^{-1} r(f,a,\varepsilon_n x_n)\}$. Then, $K \in \mathcal{K}_s$ and

$$f'(a)(x_n) \in M_2 + K \in \mathcal{M}_2 .$$

Thus, $f'(a)$ is (M_1,M_2)-preserving. To prove the converse, let $M_1 \in \mathcal{M}_1$, $\{\varepsilon_n\} \in (c_0)$ and $\{x_n\} \subset M_1$. Then, since $f'(a)$ is (M_1,M_2)-preserving, for the K defined above, we have

$$\varepsilon_n^{-1} [f(a+\varepsilon_n x_n) - f(a)] \in M_2 + K \in \mathcal{M}_2 .$$

Hence, f is (M_1,M_2)-preserving at a.

Not every (M_1,M_2)-preserving mapping is $\mathcal{M}_1$-differentiable. In fact, we need rather heavy restrictions to have differentiability.

(3.2.3) <u>Let</u> $f : A \to f(A)$ <u>be a homeomorphism and</u> (M_1,M_2)-<u>preserving at</u> a. <u>If the inverse mapping</u> $f^{-1} : f(A) \to A$ <u>is</u> $\mathcal{M}_2$-<u>differentiable at</u> $b = f(a)$ <u>and</u> $(f^{-1})'(b) \in \text{Isom}(F,E)$, <u>then</u> f <u>is</u> $\mathcal{M}_1$-<u>differentiable at</u> a.

PROOF. Let $g = f^{-1}$, $M_1 \in \mathcal{M}_1$ and $x \in M_1$. Then,

$$f(a+x) - f(a) - g'(b)^{-1}(x) = g'(b)^{-1}[r(g,b,f(a+x) - f(a)] .$$

If $f \notin D_{M_1}(a,F)$, there exist $M_1 \in \mathcal{M}_1$, $\{\varepsilon_n\} \in (c_0)$ and $\{x_n\} \subset M_1$ such that $\varepsilon_n^{-1} r(g,b,f(a+\varepsilon_n x_n) - f(a)) \not\to 0$. However, by the assumption, there is $M_2 \in \mathcal{M}_2$ such that

$$\varepsilon_n^{-1}[f(a+\varepsilon_n x_n) - f(a)] \in M_2 ,$$

which contradicts the fact that $g \in D_{M_2}(b,E)$.

This is equivalent to the following theorem on the differentiability of inverse mappings.

(3.2.4) Let $f : A \to f(A)$ be a homeomorphism, M_1-differentiable at a and $f'(a) \in \text{Isom}(E,F)$. Then, if the inverse f^{-1} is (M_2, M_1)-preserving at $b = f(a)$, then f^{-1} is M_2-differentiable at b and $(f^{-1})'(b) = f'(a)^{-1}$.

If the spaces involved are normed, the condition on f^{-1} can be simplified.

(3.2.5) Let $E, F \in \text{NLS}$. Let $f : A \to f(A)$ be a homeomorphism, differentiable at a and $f'(a) \in \text{Isom}(E,F)$. Then, f^{-1} is differentiable at $b = f(a)$ and $(f^{-1})'(b) = f'(a)^{-1}$.

PROOF. Since $f'(a)$ is invertible, there exists $\lambda > 0$ such that

$$\|f'(a)(x)\| \geq \lambda \|x\| \quad \text{for all} \quad x \in E .$$

Since $f \in D(a,F)$, there exists $\alpha > 0$ such that

$$\|f(a+x) - f(a) - f'(a)(x)\| \leq \tfrac{1}{2}\lambda \|x\| \quad \text{if} \quad \|x\| < \alpha$$

Since $g = f^{-1}$ is continuous at b, there exists $\beta > 0$ such that

$$\|g(b+y) - g(b)\| < \alpha \quad \text{if} \quad \|y\| < \beta$$

Therefore, if $\|y\| < \beta$,

$$\|f(g(b+y)) - f(a) - f'(a)(g(b+y) - g(b)\| \leq \tfrac{1}{2}\lambda \|g(b+y) - g(b)\|$$

which implies

$$\|f'(a)(g(b+y) - g(b))\| - y \leq \tfrac{1}{2}\lambda \|g(b+y) - g(b)\| ,$$

or,

$$\tfrac{1}{2}\lambda \|g(b+y) - g(b)\| \leq \|y\| .$$

Therefore, g is $(\mathcal{B}(F), \mathcal{B}(E))$-preserving at b, and, hence, by (3.2.4), $g \in D(b,E)$ and $g'(b) = f'(a)^{-1}$.

In this theorem, the assumption $E, F \in NLS$ can not be replaced by MLS, and, even for NLS, the differentiability can not be replaced by Hadamard differentiability.

1°. Let $E = F = R^\infty$, the product space of countable R's. Put

$$a_k = (2^{-2^k}, 2^{-2^{k-1}}, \ldots, 2^{-1}, 0, \ldots),$$

$b_k = (0, \ldots, 0, 1, 0, \ldots)$ with 1 at the k-th coordinate,

$$a_{k,n} = (2 - \frac{1}{n}) a_k \quad \text{and} \quad b_{k,n} = (2 - \frac{1}{n}) b_k .$$

Define $f : E \to E$ by

$$f(a_{k,n}) = a_{k,n-1} \quad (k=1,2,\ldots; \; n=2,3,\ldots)$$

$$f(a_k) = b_k \;(k=1,2,\ldots), \quad f(b_{k,n}) = b_{k,n+1} \;(k=1,2,\ldots; \; n=1,2,\ldots)$$

$$f(x) = x \quad \text{if} \quad x \notin \{a_{k,n}\} \cup \{b_{k,n}\}.$$

Then, f satisfies the assumptions of (3.2.5) at $a = 0$ with $A = E$ and $f'(0) = 1$. However, f^{-1} is not differentiable at 0.

2°. Let $E = F = \ell^2$ and e_k be the k-th coordinate element. Put

$$\lambda_k = 1/k, \quad \mu_k = 1/k^2, \quad \lambda_{k,n} = (2 - \frac{1}{n}) \lambda_k \quad \text{and} \quad \mu_{k,n} = (2 - \frac{1}{n})^{-1} \mu_k .$$

Define $f : E \to E$ by

$$f(\lambda_{k,n} e_k) = \lambda_{k,n-1} e_k \quad (k=1,2,\ldots; \; n=2,3,\ldots)$$

$$f(\lambda_k e_k) = \mu_k e_k \quad (k=1,2,\ldots)$$

$$f(\mu_{k,n} e_k) = \mu_{k,n+1} e_k \quad (n=1,2,\ldots; \; n=1,2,\ldots)$$

$$f(x) = x \quad \text{if} \quad x \notin \{\lambda_{k,n} e_k\} \cup \{\mu_{k,n} e_k\} .$$

Then, $f : E \to E$ is a bijection, f^{-1} is continuous at 0 and $f \in D_H(0, E)$ with $f'(0) = 1$. However, $f^{-1} \notin D_H(0, E)$.

Finally, we give the C^n-version of (3.2.5).

(3.2.6) <u>Let</u> $E, F \in$ NLS <u>and</u> $f : A \to f(A)$ <u>be a homeomorphism.</u> <u>If</u> $f \in C^n(A, F)$ <u>and</u> $f'(x) \in$ Isom(E,F) <u>for all</u> $x \in A$, <u>then</u> $f^{-1} \in C^n(f(A), E)$.

PROOF. For $n = 1$, it follows from (3.2.5) that $f^{-1} \in D(f(A), E)$ and

$$(f^{-1})'(y) = f'(f^{-1}(y))^{-1} = (\text{inv} \circ f' \circ f^{-1})(y) \quad \text{for} \quad y \in f(A).$$

Since all mappings involved are continuous, $f^{-1} \in C^1(f(A), E)$.

To prove the general case, suppose that the theorem holds for $n = k$, and assume that $f \in C^{k+1}(A, F)$. Then, the k-th derivative of $(\text{inv} \circ f' \circ f^{-1})(y)$ contains $(\text{inv})^{(k)}$, $f^{(k+1)}$, $(f^{-1})^{(k)}$, which all exist and continuous. Therefore, $(f^{-1})'(y)$ is k-times continuously differentiable, or, $f^{-1} \in C^{k+1}(f(A), E)$.

An LCS-version of this theorem will be given in §3.4.

NOTE: Except for minor changes, all materials in this section were taken from Averbukh-Smolyanov [1,2]. The assumptions of (3.2.5) are too strong. In fact, the assumption that "$f : A \to f(A)$ is a homeomorphism" can be replaced by "$f : A \to f(A)$ is a bijection and f^{-1} is continuous at $f(a)$". The proof remains unchanged.

Suhinin [1] has proved the following two facts. Let $E \in$ LCS and $f : E \to E$. Assume that there are open subsets U and V such that $x_0 \in U$, $y_0 = f(x_0) \in V$, $f : V \to V$ is a bijection and $g = f^{-1} : V \to U$ is continuous at y_0.

(1) Assume that E is a **strict** inductive limit of $E_n \in$ BS $(n=1,2,\ldots)$, f is Fréchet differentiable at x_0 and $f'(x_0) \in$ Isom(E,E). Then, g is Fréchet differentiable at y_0 and $g'(y_0) = [f'(x_0)]^{-1}$.

(2) If f is strongly Fréchet differentiable at x_0 and g is continuous on V, then g is strongly Fréchet differentiable at y_0 and $g'(y_0) = [f'(x_0)]^{-1}$.

§3.3 The space $L_p(E,F)$

As the preparation for an LCS-version of the inverse mapping theorem, this section is devoted to the study of the p-bounded linear mappings that have been introduced by Garnir-de Wilde-Schmets [1], who called them completely bounded operators.

In this section, we assume that $E,F \in$ LCS and $p \in P(E)$.

[3.3.1] $u \in L(E,F)$ is said to be p-bounded if, for any $q \in P(F)$, there exists $\lambda_q > 0$ such that

$$q(u(x)) \leq \lambda_q p(x) \quad \text{for all} \quad x \in E .$$

The set of all p-bounded linear mappings of E into F is obviously a linear subset of $L(E,F)$, and will be denoted by $L_p(E,F)$.

By the Hahn-Banach theorem, there exists $\bar{a} \in \bar{E}$ such that $\bar{a} \neq 0$ and $|<x,\bar{a}>| \leq p(x)$ for all $x \in E$. Then, for any $q \in P(F)$ and $a \in E$,

$$q((a \otimes \bar{a})(x)) \leq q(a).p(x) \quad \text{for all} \quad x \in E,$$

which shows that $a \otimes \bar{a} \in L_p(E,F)$. Hence, $L_p(E,F)$ is not trivial.

For $u \in L_p(E,F)$ and $q \in P(F)$, we put

$$\hat{q}(u) = \sup_{p(x) \neq 0} q(u(x))/p(x) .$$

Then, $\hat{q}(u)$ is finite and

1°. if $\hat{q}(u) = 0$ for all $q \in P(F)$, then $u = 0$;
2°. $\hat{q}(\alpha u) = |\alpha|\hat{q}(u)$ if $\alpha \in R$ and $u \in L_p(E,F)$;
3°. $\hat{q}(u+v) \leq \hat{q}(u) + \hat{q}(v)$ if $u,v \in L_p(E,F)$;
4°. $q(u(x)) \leq \hat{q}(u)p(x)$ if $x \in E$ and $u \in L_p(E,F)$.

The following theorems have been proved in Garnir- de Wilde-Schmets [1].

(3.3.2) __If__ $u \in L_p(E,F)$ __and__ $v \in L(F,G)$, __then__ $v \circ u \in L_p(E,G)$.

PROOF. For any $q \in P(G)$, choose $q_1 \in P(F)$ such that

$$q(v(x)) \leq \lambda q_1(x) \quad \text{for some} \quad \lambda > 0 \quad \text{and all} \quad x \in F.$$

Then, $q(v(u(x))) \leq \lambda \hat{q}_1(u) p(x)$ for all $x \in E$, which implies $v \circ u \in L_p(E,G)$.

(3.3.3) __If__ $L_p(E,F)$ __contains an invertible element of__ $L(E,F)$, __then__, E __and__ F __are normable__.

PROOF. Let $u \in L_p(E,F)$ be invertible. Then, $p \circ u^{-1}$ is a norm on F and p is a norm on E.

(3.3.4) __Let__ E __be sequentially complete__, $u \in L_p(E,E)$ __and__ $\hat{p}(u) < 1$. __Then__, $1 - u$ __has the inverse and it is expressed as__

$$(1-u)^{-1} = \sum_{n=0}^{\infty} u^n,$$

__which is absolutely convergent in__ $L(E,E)$. __Moreover, for any__ $q \in P(E)$ __and__ $B \in \mathcal{B}(E)$,

$$\sup_{x \in B} q((1-u)^{-1}(x)) \leq \sup_{x \in B} q(x) + \frac{\hat{q}(u)}{1-\hat{p}(u)} \sup_{x \in B} p(x).$$

PROOF. First, we prove that the series is convergent at any point $x \in E$. Since E is sequentially complete, we have only to show that $\{\sum_{n=0}^{k} u^n(x)\}_{k=1,2,\ldots}$ is a Cauchy sequence. Now, let $q \in P(E)$. Since $q(u^n(x)) \leq \hat{q}(u) p(u^{n-1}(x)) \leq \hat{q}(u) \hat{p}(u)^{n-1} p(x)$, $q(\sum_{n=m}^{k} u^n(x)) \leq \hat{q}(u) \sum_{n=m}^{k} p(u)^{n-1} p(x) \to 0$ if $k,m \to \infty$. Thus, we can define $v : E \to E$ by

$$v(x) = \sum_{n=0}^{\infty} u^n(x) \quad \text{for all} \quad x \in E.$$

Obviously, it is linear and, for any $q \in P(E)$,

$$q(v(x)) \leq \sum_{n=0}^{\infty} q(u^n(x)) \leq q(x) + (\hat{q}(u) \sum_{n=0}^{\infty} \hat{p}(u)^n) p(x) ,$$

which implies that v is continuous.

To prove that v is the sum in $L(E,E)$, take $q \in P(E)$ and $B \in \mathcal{B}(E)$. Then,

$$\sup_{x \in B} q(v(x) - \sum_{n=0}^{k} u^n(x)) \leq \sum_{n=k+1}^{\infty} \sup_{x \in B} q(u^n(x))$$

$$\leq \hat{q}(u) (\sum_{n=k+1}^{\infty} \hat{p}(u)^{n-1}) \sup_{x \in B} p(x) \to 0 \quad \text{if} \quad k \to \infty .$$

Thus, $v = \sum_{n=0}^{\infty} u^n$ in $L(E,E)$.

Finally, to show that $v = (1-u)^{-1}$, take $q \in P(E)$. Since $1 - u \in L(E,E)$, there exists $\lambda > 0$ such that, for some $q_1 \in P(E)$,

$$q((1-u)(x)) \leq \lambda q_1(x) \quad \text{for all} \quad x \in E .$$

Then,

$$q((1-u)v(x)-x) \leq q((1-u) \sum_{n=0}^{k} u^n(x)-x) + q((1-u)(v(x) - \sum_{n=0}^{k} u^n(x)))$$

$$\leq q(u^{k+1}(x)) + \lambda q_1(v(x) - \sum_{n=0}^{k} u^n(x)) \to 0 \quad \text{if} \quad k \to \infty,$$

which implies $(1-u)v = 1$. Similarly, we have $v(1-u) = 1$.

We add two theorems for spaces equipped with projective and inductive topologies.

(3.3.5) **Assume that** F **is equipped with the projective topology determined by the family** $\{F_\sigma, \pi_\sigma\}$ (See §1.5 for the definition). **Then, for** $p \in P(E)$, $u \in L_p(E,F)$ **if and only if** $\pi_\sigma \circ u \in L_p(E,F_\sigma)$ **for all** σ.

PROOF. Since the necessity follows from (3.3.2), let us assume that $\pi_\sigma \circ u \in L_p(E,F_\sigma)$ for all σ. Then, $u \in L(E,F)$ and for any $q_\sigma \in P(F_\sigma)$ there exists $\lambda_\sigma > 0$ such that

$$(q_\sigma \circ \pi_\sigma)(u(x)) = q_\sigma((\pi_\sigma \circ u)(x)) \leq \lambda_\sigma p(x) \quad \text{for all} \quad x \in E.$$

On the other hand, for any $q \in P(F)$, there exist finite $q_\sigma \in P(F_\sigma)$ such that $q = \Sigma \, q_\sigma \circ \pi_\sigma$. Therefore,

$$q(u(x)) \leq (\Sigma \lambda_\sigma) p(x) \quad \text{for all} \quad x \in E,$$

which shows that $u \in L_p(E,F)$.

(3.3.6) <u>Assume that E is equipped with the inductive topology determined by the family $\{E_\sigma, i_\sigma\}$, in other words, we assume that the conditions 1, 2 and 3 of §1.6 are satisfied. Then, for $p \in P(E)$, $u \in L_p(E,F)$ if and only if $u \circ i_\sigma \in L_{p \circ i_\sigma}(E_\sigma, F)$ for all σ.</u>

PROOF. We use the following facts: $E = \bigcup_\sigma i_\sigma(E_\sigma)$ and

$$P(E) = \{\text{semi-norm } p_1 \text{ on } E : p_1 \circ i_\sigma \in P(E_\sigma)\}.$$

Since the necessity is obvious, we assume that $u \circ i_\sigma \in L_{p \circ i_\sigma}(E_\sigma, F)$ for all σ. Then, $u \in L(E,F)$ and, for any $q \in P(F)$, there exists $\lambda_q > 0$ such that

$$q((u \circ i_\sigma)(x)) \leq \lambda_q (p \circ i_\sigma)(x) \quad \text{for all} \quad x \in E_\sigma.$$

Now, for any $x \in E$, take σ such that $x = i_\sigma(x_\sigma)$ for some $x_\sigma \in E_\sigma$. Then,

$$q(u(x)) = q((u \circ i_\sigma)(x_\sigma)) \leq \lambda_q p(x),$$

which implies that $u \in L_p(E,F)$.

§3.4 C_p-<u>mappings and an inverse mapping theorem</u>

Let $E, F \in LCS$, $a \in A \in \mathcal{O}(E)$ and $f : A \to F$.

The following definition is introduced solely to overcome the difficulties which appear in the process of transplanting the NLS proof of the inverse mapping theorem to TLS.

[3.4.1] Let $f \in D(A,F)$ and $p \in P(E)$. Then, f is said to **be** C_p^1 **at** a if, for any $\varepsilon > 0$ and $q \in P(F)$, there exists $\delta > 0$ such that

1°. if $P(x) \leq \delta$, then $a + x \in A$;

2°. if $p(x) \leq \delta$, then $f'(a+x) - f'(a) \in L_p(E,F)$ and $\hat{q}[f'(a+x) - f'(a)] \leq \varepsilon$;

3°. $\sup_{p(x) \leq \delta} \hat{q}_1[f'(a+x) - f'(a)] < \infty$ for any $q_1 \in P(F)$.

It is obvious that any element of $L(E,F)$ is C_p^1 everywhere for any $p \in P(E)$, and, if $E, F \in NLS$, this notion coincides with the usual continuous differentiability.

(3.4.2) **If** f **is** C_p^1 **at** a, f **is continuous there**.

PROOF. Let $\{x_\sigma\}$ be a net convergent to zero. Then, by the definition, for any $\varepsilon > 0$ and $q \in P(F)$, there exists σ_0 such that $\sigma \geq \sigma_0$ implies

$$\hat{q}[f'(a+\theta x_\sigma) - f'(a)] \leq \varepsilon \quad \text{for any } \theta \in [0,1].$$

On the other hand, by the mean value theorem,

$$q(f(a+x_\sigma) - f(a)) \leq q(f'(a+\theta x_\sigma)(x_\sigma)$$

$$\leq q[f'(a+\theta x_\sigma)(x_\sigma) - f'(a)(x_\sigma)] + q(f'(a)(x_\sigma))$$

$$\leq \varepsilon p(x_\sigma) + q(f'(a)(x_\sigma)) \to 0,$$

which shows the continuity of f at a.

(3.4.3) **If** f **is** C_p^1 **at** a, **then** $u \circ f : E \to E$ **is** C_p^1 **at** a **for any** $u \in L(F,E)$.

PROOF. For any $q \in P(E)$, $q \circ u \in P(F)$ if $u \in L(F,E)$. Moreover,

$$\widehat{q \circ u}[f'(a+x) - f'(a)] = \hat{q}[(u \circ f)'(a+x) - (u \circ f)'(a)].$$

Now, by the assumption, for any $\epsilon > 0$ and $q \circ u$, there exists $\delta > 0$ such that the three conditions of [3.4.1] are satisfied. Then, if $p(x) \leq \delta$, it follows from (3.3.2) that

$$(u \circ f)'(a+x) - (u \circ f)'(a) = u(f'(a+x) - f'(a)) \in L_p(E,F),$$

and other conditions are satisfied for q and $u \circ f$. Thus, $u \circ f$ is C_p^1 at a.

The following theorem generalizes (3.2.5) for $n = 1$.

(3.4.4) Let $f : A \to f(A)$ be a homeomorphism and $f \in D(A,F)$.

1°. If $f'(a) \in \text{Isom}(E,F)$ and f is C_p^1 at a, then $f^{-1} \in D(f(a),E)$.

2°. Let E be sequentially complete. If $f'(x) \in \text{Isom}(E,F)$ for all $x \in A$ and f is C_p^1 at every point of A, then $f^{-1} \in C^1(f(A),E)$.

PROOF. 1°. By (3.4.3), we can assume that $E = F$ and $f'(a) = 1$. By (3.2.4), we need to show that f^{-1} is $(\mathcal{B}(E), \mathcal{B}(E))$-preserving at $b = f(a)$. Therefore, we assume $B \in \mathcal{B}(E)$, $\{\epsilon_n\} \in (c_0)$ and $\{y_n\} \subset B$, and we shall show that

$$\epsilon_n^{-1}[f^{-1}(b+\epsilon_n y_n) - f^{-1}(b)]$$

is a bounded sequence. Let us put

$$x_n = \epsilon_n^{-1}[f^{-1}(b+\epsilon_n y_n) - f^{-1}(b)] \quad \text{and} \quad g = 1 - f.$$

Then,

$$x_n = y_n + \epsilon_n^{-1}[g(a+\epsilon_n x_n) - g(a)],$$

and, since $\epsilon_n y_n \to 0$ and f is a homeomorphism, we have $\epsilon_n x_n \to 0$. Now, since f is C_p^1 at a, there exists $\delta > 0$ such that, if $p(x) \leq \delta$,

$$a + x \in A, \ g'(a+x) \in L_p(E,E), \ \hat{p}(g'(a+x)) \leq \frac{1}{2}$$

and $\sup_{p(x) \leq \delta} \hat{q}(g'(a+x)) < \infty$ for any $q \in P(E)$.

Then, if $p(\varepsilon_n x_n) \leq \varepsilon$, which is true for almost all n,

$$p(x_n) \leq p(y_n) + p(\varepsilon_n^{-1}[g(a+\varepsilon_n x_n) - g(a)])$$

$$\leq p(y_n) + p(g'(a+\theta_n \varepsilon_n x_n)(x_n)) \quad \text{for some } \theta_n \in (0,1)$$

$$\leq p(y_n) + \frac{1}{2}p(x_n),$$

which leads to

$$p(x_n) \leq 2p(y_n).$$

Then, by the same argument, for any $q \in P(E)$, we have

$$q(x_n) \leq q(y_n) + \sup_{p(x)\leq\delta} \hat{q}[g'(a+x)]p(y_n).$$

Hence, $\{x_n\} \in B(E)$.

$2°$. We show that f^{-1} is C^1 at $f(a)$. We can again assume that $E = F$ and $f'(a) = 1$. Let $y_\sigma \to 0$ and put

$$x_\sigma = f^{-1}(b+y_\sigma) - f^{-1}(b) \quad \text{for } b = f(a).$$

Since f^{-1} is continuous, $x_\sigma \to 0$ and $a + x_\sigma = f^{-1}(b+y_\sigma)$. Therefore, what we have to show is

$$f'(a+x_\sigma)^{-1} = (f^{-1})'(b+y_\sigma) \to 0.$$

Now, since f is C_p^1 at a, there exists σ_0 such that $\sigma \geq \sigma_0$ implies

$$\hat{p}[f'(a+x_\sigma) - 1] < 1.$$

Therefore, by (3.3.4),

$$f'(a+x_\sigma)^{-1} = \sum_{n=0}^{\infty} (1-f'(a+x_\sigma))^n,$$

and, hence, for any $q \in P(E)$,

$$q((f'(a+x_\sigma)^{-1}(z)) \leq \hat{q}[1-f'(a+x_\sigma)] \cdot \frac{p(z)}{1-\hat{p}[1-f'(a+x_\sigma)]}.$$

From the definition [3.4.1]. 2°, we have

$$\hat{q}[1-f'(a+x_\sigma)] \to 0.$$

Therefore, for any $B \in \mathcal{B}(E)$,

$$\sup_{z \in B} q((f'(a+x_\sigma))^{-1}(z)) \to 0,$$

which shows that f^{-1} is C^{-1} at $f(a)$.

The following theorem is an LCS-version of the inverse function theorem.

(3.4.5) Let $E, F \in$ LCS and E be sequentially complete. Assume that $a \in A \in \mathcal{O}(E)$ and $f \in D(A,F)$. If f is C^1_p at each point of A and $f'(a) \in \text{Isom}(E,F)$, then f is a local homeomorphism at a and f^{-1} is C^1 at $f(a)$.

PROOF. Again, we can assume that $E = F$ and $f'(a) = 1$. Moreover, we can assume that $a = 0$.

Since f is C^1_p at a, there exists $\delta_0 > 0$ such that $p(x) \leq \delta_0$ implies $x \in A$ and

$$\hat{p}[1-f'(x)] \leq \frac{1}{2} \quad \text{and} \quad \lambda_q = \sup_{p(x) \leq \delta_0} \hat{q}[1-f'(x)] < +\infty$$

for each $q \in P(E)$. Put

$$g = 1 - f \quad \text{and} \quad V = \{x \in E : p(x) < \frac{1}{2}\delta_0\}.$$

Since p is continuous, V is open and

$$\bar{V} = \{x \in E : p(x) \leq \frac{1}{2}\delta_0\} \subset \frac{1}{2}A.$$

Then, we have

$$p(g(x) - g(y)) \leq \frac{1}{2}p(x-y) \quad \text{if} \quad x,y \in 2\bar{V}, \tag{1}$$

because by the mean value theorem,

$$p(g(x) - g(y)) \leq p(g'(y+\theta(x-y))(x-y))$$

$$\leq \hat{p}[g'(y+\theta(x-y))]p(x-y) \leq \frac{1}{2}p(x-y).$$

Similarly, for any $q \in P(E)$, we have

$$q(x-y) \leq q(f(x) - f(y)) + 2\lambda_q p(f(x) - f(y)) \quad \text{if} \quad x,y \in 2\bar{V}, \tag{2}$$

because we have

$$q(x-y) \leq q(f(x) - f(y)) + q(g(x) - g(y))$$

$$\leq q(f(x) - f(y)) + \lambda_q p(x-y),$$

in which $p(x-y)$ can be replaced by $2p(f(x) - f(y))$, since it follows from (1) that

$$p(x-y) \leq p(f(x) - f(y)) + p(g(x) - g(y))$$

$$\leq p(f(x) - f(y)) + \frac{1}{2}p(x-y).$$

Now, we shall prove that "<u>for any</u> $y \in \bar{V}$ <u>there exists</u> $x \in 2\bar{V}$ <u>uniquely such that</u> $y = f(x)$". To do this, let us consider the mapping g_y defined by

$$g_y(x) = y + g(x) \quad \text{for} \quad x \in 2\bar{V}.$$

Since, by (1),

$$p(g_y(x)) \leq p(y) + p(g(x)) \leq p(y) + \frac{1}{2}p(x) \leq \delta_0,$$

g_y maps $2\bar{V}$ into itself. For $y_1 = g_y(y)$,

$$p(y_1-y) = p(g_y(y) - g_y(0)) = p(g(y)) \leq \frac{1}{2}p(y) \leq \frac{1}{2^2}\delta_0.$$

For $y_2 = g_y(y_1)$, the same argument implies $p(y_1-y_2) \leq \delta_0/2^3$. Continuing this process, we have

$$p(y_n-y_{n-1}) \leq \delta_0/2^{n+1} \quad \text{for} \quad y_n = g_y(y_{n-1}).$$

Thus,
$$p(y_{n+m}-y_n) \leq \delta_0/2^{n+1}.$$

Moreover, for any $q \in P(E)$,
$$q(y_{n+m}-y_n) \leq q(g(y_{n+m-1}) - g(y_{n-1})) \leq \lambda_q \delta_0/2^n.$$

Now, this means that $\{y_n\}$ is a Cauchy sequence and, hence, it converges to some point $x \in 2\bar{V}$. Since f is continuous in A by (3.4.2), we have $y = f(x)$. The uniqueness of such x follows from (2), which also implies that $y \in V$ implies $x \in 2V$. Let us put
$$U = f^{-1}(V) \cap 2V.$$

Then, U is an open set containing 0, and, since $2V \subset A$, we have $U \subset A$. Since $f : U \to V$ is a bijection by what has just been proved and $f^{-1} : V \to U$ is continuous by (2), f is a homeomorphism of U onto V.

Finally, we shall prove that f^{-1} is continuously differentiable on V. By (3.4.4), we have to show that the following two conditions are satisfied:

1°. $f'(x) \in \text{Isom}(E,E)$ if $x \in U$;

2°. $f : U \to V$ is C_p^1 at each point of U.

The first condition is satisfied because of (3.3.4) and the fact that $\hat{p}[1-f'(x)] \leq \frac{1}{2}$ if $x \in U$.

Since $f : A \to F$ is C_p^1 at every point of A, to show that the second condition is satisfied we only have to show that for any $x_0 \in U$ there exists $\delta > 0$ such that $p(x) \leq \delta$ implies $x_0 + x \in U$.

Now, since $x_0 \in 2V$ and $f(x_0) \in V$, we can choose $\delta > 0$ such that
$$\delta < \text{Min.} \{\delta_0 - p(x_0), \frac{2}{3}(\frac{\delta_0}{2} - p(f(x_0)))\}.$$

Assume that $p(x) \leq \delta$. Then,
$$p(x_0+x) \leq p(x_0) + p(x) < \delta_0,$$

which implies that

$$\hat{p}[g'(x_0+\theta x)] \le \frac{1}{2} \quad \text{for any } \theta \in (0,1).$$

Therefore, by the mean value theorem,

$$p(f(x_0+x)) \le p(x) + p(g(x_0+x) - g(x_0)) + p(f(x_0))$$

$$\le \delta + \hat{p}[g'(x_0+\theta x)]p(x) + p(f(x_0))$$

$$\le \frac{3}{2}\delta + p(f(x_0)) \le \delta_0/2,$$

which implies $x_0 + x \in f^{-1}(V)$. Hence, $x_0 + x \in U$. Therefore, $f^{-1} : V \to U$ is continuously differentiable.

When the spaces involved are normed, we have the usual "inverse mapping theorem".

(3.4.6) <u>Let</u> $E, F \in BS$, $A \in \mathcal{O}(E)$ <u>and</u> $f \in C^n(A,F)$. <u>Assume that, for some point</u> $a \in A$, $f'(a) \in \text{Isom}(E,F)$. <u>Then</u>, f <u>is a local</u> C^n-<u>isomorphism at</u> a.

PROOF. We need only to show that $f^{-1} \in C^n(V,U)$ for the sets U and V determined in the above proof. However, this is an immediate consequence of (3.2.6).

NOTE: When the space E and F are general LCS's, it is impossible to obtain a faithful generalization of (3.4.6). The following two examples show that, even in separable complete MLS's, there are C^∞-mappings whose derivative at a point is invertible but still not one-to-one :

1°. (Keller [1] and Averbukh-Smolyanov [2]). Let R^N be the product of countable number of R. Then, the mapping $f : R^N \to R^N$ defined by

$$f : (\xi_n) \mapsto (\xi_n - \xi_n^2)$$

is C^1 everywhere and $f'(0) = 1$. However, there is no neighbourhood of zero in which f is one-to-one. For instance, take $x_n = (\delta_{n,k})_{k \in N}$. Then, $x_n \to 0$

and $f(x_n) = 0$.

2°. (Eells [1]). Let E be the set of all continuous functions on R with the topology of compact convergence. Then, the mapping $\exp : E \to E$ is C^1 everywhere and $(\exp)'(0) = 1$. However, no neighbourhood of $1 = \exp(0)$ is the image under exp of a neighbourhood of zero.

§3.5 Other theorems on inverse mappings

When the spaces are BS's, there are a number of theorems which are related to (3.4.6). The first three theorems in this section are concerned with the case when $f'(a)$ is not invertible but only injective or surjective.

Throughout, we assume that $E, F \in BS$, $a \in A \in \mathcal{O}(E)$ and $f : A \to F$.

(3.5.1) (Local injection theorem) Assume that $f \in C^1(A,F)$. If $f'(a)$ has a continuous inverse (it does not need to be onto), then there is $U \in \mathcal{O}(E)$ such that $a \in U \subseteq A$ and f is injective on U. The inverse f^{-1} defined on $f(U)$ is continuous. However, f(U) is not necessarily an open set.

PROOF. From the assumption, there are $\alpha > 0$ and $\beta > 0$ such that

$$\|f'(a)(x)\| \geq \alpha \|x\| \quad \text{for every} \quad x \in E$$

and

$$\|f(x_1) - f(x_2) - f'(a)(x_1-x_2)\| \leq \frac{\alpha}{2}\|x_1-x_2\| \quad \text{if} \quad \|x_i-a\| < \beta \quad (i=1,2).$$

Hence, if $x_i \in B(a,\beta)$ (i=1,2),

$$\alpha\|x_1-x_2\| \leq \|f'(a)(x_1-x_2)\| \leq \|f(x_1) - f(x_2)\| + \frac{\alpha}{2}\|x_1-x_2\|.$$

Therefore,

$$\frac{\alpha}{2}\|x_1-x_2\| \leq \|f(x_1) - f(x_2)\|,$$

from which it follows that f is injective on $B(a,\beta)$ and the inverse mapping is continuous on $f(B(a,\beta))$. The fact that f(U) may not be open can be easily seen if we take the special case when f is linear.

(3.5.2) (Local surjection theorem)   Assume that $f \in C^1(A,F)$. If $f'(a)$ is surjective, then there are $V \in \mathcal{O}(F)$ such that $f(a) \in V$ and $U \in \mathcal{O}(E)$ such that $a \in U \subset A$ for which we have $V \subset f(U)$.

PROOF.   By considering $f(a+x) - f(a)$ instead of $f(a)$, we can assume that $a = 0$ and $f(0) = 0$. Since $u = f'(0)$ is surjective, there is $\alpha > 0$ such that for any $y \in F$ there is $x \in E$ such that $\alpha\|x\| \leq \|y\|$ and $y = u(x)$ (see, for instance, Taylor [1], p.170). Since f is C^1 at a, there exists $\beta > 0$ such that $B(0,\beta) \subset A$ and

$$\|f(x_1) - f(x_2) - f'(a)(x_1-x_2)\| \leq \tfrac{\alpha}{2}\|x_1-x_2\| \quad \text{if} \quad x_i \in B(0,\beta) \ (i=1,2).$$

For $\gamma = \alpha\beta/2$, we put $V = B(0,\gamma)$, and we take $y \in V$. We shall find $x \in U = B(0,\beta)$ such that $y = f(x)$.

Now, put $x_0 = 0$ and $y_0 = y$. By induction, we determine sequences $\{x_k\}$ and $\{y_k\}$ such that

1°. $\|x_k - x_{k-1}\| \leq \alpha^{-1}\|y_{k-1}\| \leq \|y\|/2^{k-1}\alpha$;

2° $y_{k-1} = u(x_k - x_{k-1})$;

3°. $y_k = y_{k-1} + f(x_{k-1}) - f(x_k)$.

To do this for $k = 1$, we choose x_1 such that

$$\alpha\|x_1\| \leq \|y\| \quad \text{and} \quad y = u(x_1).$$

Then, $\|x_1\| \leq \alpha^{-1}\|y\| < \beta/2$, or $x_1 \in U$. Put

$$y_1 = y_0 + f(x_0) - f(x_1).$$

These elements x_1 and y_1 satisfy the conditions 1°, 2° and 3°. Assume that we have constructed $\{x_k\}$ and $\{y_k\}$ for $k = 1,2,\ldots,n$. Then, we take x_{n+1} such that

$$y_n = u(x_{n+1}-x_n) \quad \text{and} \quad \alpha\|x_{n+1}-x_n\| \leq \|y_n\|.$$

Then, by 1°,

$$\|x_{n+1}\| \leq \|x_{n+1}-x_n\| + \|x_n-x_{n-1}\| + \ldots + \|x_1-x_0\|$$

$$\leq \alpha^{-1}(\sum_{k=1}^{n+1} \frac{1}{2^{k-1}})\|y\| < \alpha^{-1}(2 - \frac{1}{2^n})\cdot\frac{\alpha\beta}{2} < \beta,$$

or, $x_{n+1} \in U$, and put

$$y_{n+1} = y_n + f(x_n) - f(x_{n+1}).$$

Then, x_{n+1} and y_{n+1} satisfy the three conditions for $k = n + 1$. Since $\{x_k\}$ is a Cauchy sequence, $x_k \to x_0$ for some x_0 and

$$\|x_0\| = \lim_{k\to\infty} \|x_{k+1}\| \leq \lim_{n\to\infty} \alpha^{-1}(\sum_{k=1}^{n+1} \frac{1}{2^{k+1}})\|y\| = 2\alpha^{-1}\|y\| < \beta,$$

which means that $x_0 \in U$.

On the other hand, it follows from 1° that $y_k \to 0$. Now, add up 3° for $k = 1,2,\ldots,n$. Then, $y_n = y - f(x_n)$. Hence, $y = f(x_0)$.

The following theorem is an immediate corollary of (3.5.2).

(3.5.3) (Open mapping theorem) <u>Assume that</u> $f \in C^1(A,F)$. <u>If</u> $f'(x)$ <u>is onto for any</u> $x \in A$, <u>then</u> $f(A)$ <u>is open</u>.

In the following two theorems, which correspond to (3.5.1) and (3.5.2), stronger assumptions are made on $f'(a)$, namely, in addition to its being injective or surjective, it is supposed to split the space so that it in fact becomes an isomorphism of E onto a subspace of F or of a subspace of E onto F. Naturally, under these assumptions, we obtain splitted versions of the inverse mapping theorem (3.4.6).

(3.5.4) (Split C^n-imbedding theorem) <u>Assume that</u> $F_1, F_2 \in BS$, $f \in C^n(A, F_1 \times F_2)$ <u>and</u> $f(a) = (0,0)$. <u>If</u> $f'(a) \in \text{Isom}(E, F_1 \times \{0\})$, <u>then, there is a local isomorphism</u> g <u>of</u> $F_1 \times F_2$ <u>at</u> 0 <u>such that</u> $g \circ f : A \to F_1 \times F_2$ <u>maps an open set</u> $U \subset A$ <u>into</u> $F_1 \times \{0\}$ <u>and induces a</u> C^n-<u>diffeomorphism of</u> U <u>on an open</u>

subset of $F_1 \times \{0\}$.

PROOF. We can assume that $E = F_1$ and $a = 0$. Consider the mapping $h : A \times F_2 \to F_1 \times F_2$ defined by

$$h(x,y_2) = f(x) + (0,y_2) \text{ for } x \in A \text{ and } y_2 \in F_2.$$

Then, $h(x,0) = f(x)$ and $h'(0,0)(x,y_2) = f'(0)(x) + (0,y_2)$. Hence, $h'(0,0) \in \text{Isom}(F_1 \times F_2, F_1 \times F_2)$ and, by (3.4.6), it is a local C^n-diffeomorphism. Let g be the inverse of h. Then, $g \circ f$ satisfies the required conditions.

(3.5.5) (Split C^n-projection theorem) <u>Assume that</u> $E = E_1 \times E_2$, $E_i \in BS$ (i=1,2), $f \in C^n(A,F)$ <u>and</u> $f(a_1,a_2) = 0$ <u>for</u> $a = (a_1,a_2) \in A$. <u>If</u> $\partial_2 f(a_1,a_2) \in \text{Isom}(E_2,F)$, <u>then there is a local isomorphism</u> g <u>of an open set</u> $U_1 \times U_2$ <u>onto an open neighbourhood of</u> a <u>contained in</u> A <u>such that the composition mapping</u>

$$U_1 \times U_2 \overset{g}{\to} A \overset{f}{\to} F$$

<u>satisfies the following condition</u> : g <u>is a projection on the second factor and</u> f <u>is an isomorphism of</u> U_2 <u>onto an open set in</u> F.

PROOF. We can assume that $(a_1,a_2) = (0,0)$ and $E_2 = F$. Consider the mapping $h : A \to E$ defined by

$$h(x_1,x_2) = (x_1, f(x_1,x_2)).$$

Then, $h'(0,0)(x_1,x_2) = (x_1, f'(0,0)(x_1,x_2))$

$$= (x_1, \partial_1 f(0,0)(x_1) + \partial_2 f(0,0)(x_2)).$$

Obviously, $h'(0,0) : E \to E$ is injective. Moreover, it is surjective, because, if $(y_1,y_2) \in E$, putting

$$x_1 = y_1 \text{ and } x_2 = \partial_2 f(0,0)^{-1}(y_2) - \partial_1 f(0,0)(x_1),$$

we have $h'(0,0)(x_1,x_2) = (y_1,y_2)$. Therefore, $h'(0,0) \in \text{Isom}(E,E)$. Hence, by (3,4,6), it has a local inverse g which satisfies the requirement.

All theorems proved above are "local" theorems, i.e. the inverses obtained only exist locally. For the existence of the global inverse, we need a considerably stronger assumption.

(3.5.6) (Hadamard-Levy theorem) If $1°$ $f \in C^1(E,F)$; $2°$ $f'(x) \in \text{Isom}(E,F)$ for all $x \in E$ and $3°$ $\sup_{x \in E} \|f'(x)^{-1}\| < \infty$, then f is a homeomorphism of E onto F and $f^{-1} \in C^1(F,E)$.

PROOF. First, we prove that f is surjective. Since $f(E)$ is open by (3,5,3), we need to show that it is closed. By (3,4,6), for each $x \in E$ there exist convex sets $U_x \subset E$ and $V_x \subset F$ such that $f : U_x \to V_x$ is a homeomorphism. It is clear that $f(\bar{U}_x) \subset \bar{V}_x$. To obtain the inverse inclusion, let $y_0 \in \bar{V}_x$. Then, $y_n \to y_0$ for some $y_n \in V_x$. Since f^{-1} is C^1 on V_x, for $x_n = f^{-1}(y_n) \in U_x$, we have

$$\|x_n - x_m\| \leq \sup_{z \in [y_n, y_m]} \|(f^{-1})'(z)\| \|y_n - y_m\|,$$

where, by the assumption $3°$,

$$\sup_{z \in [y_n, y_m]} \|(f^{-1})'(z)\| = \sup_{z \in [y_n, y_m]} \|f'(f^{-1}(z))^{-1}\| < \infty.$$

Hence, $x_n \to x_0$ for some $x_0 \in \bar{U}_x$, since E is complete, and $y_0 = f(x_0) \in f(\bar{U}_x)$. Thus, $f(\bar{U}_x) = \bar{V}_x$, and, since

$$\overline{f(E)} \subset \overline{\bigcup_{x \in E} V_x} \subset \bigcup_{x \in E} \bar{V}_x = \bigcup_{x \in E} f(\bar{U}_x) \subset f(E),$$

$f(E)$ is closed.

Next, we prove that f is injective. Let $f(x_0) = f(x_1) = y_0$ and put $x(\eta) = (1-\eta)x_0 + \eta x_1$ for $\eta \in [0,1]$ and $y(\eta) = f(x(\eta))$. Put

$$F(\xi,\eta) = (1-\xi)y_0 + \xi y(\eta) \quad \text{for} \quad \xi,\eta \in [0,1].$$

We prove that there is $G(\xi,\eta)$ such that

$$f(G(\xi,\eta)) = F(\xi,\eta) \quad \text{for all} \quad (\xi,\eta) \in [0,1] \times [0,1].$$

Now, since $F(\xi,0) = y_0 = f(x_0)$, by (3.4.6), there are open neighbourhoods U of x_0 and V of y_0 such that $f : U \to V$ is a homeomorphism. As was proved above, $f(\bar{U}) = \bar{V}$. Since $y(\eta)$ is continuous with respect to η and

$$\|F(\xi,\eta) - y_0\| = \|\xi(y(\eta) - y(0))\| \leq \|y(\eta) - y(0)\|,$$

there exists $\varepsilon > 0$ such that

$$F(\xi,\eta) \in \bar{V} \quad \text{if} \quad 0 \leq \eta \leq \varepsilon \quad \text{and} \quad 0 \leq \xi \leq 1$$

Thus, we obtain $G(\xi,\eta)$ for $(\xi,\eta) \in [0,1] \times [0,\varepsilon]$ by putting

$$G(\xi,\eta) = f^{-1}(F(\xi,\eta)).$$

Let α be the largest of such number ε. If $\alpha < 1$, then, again by (3.4.6), $G(\xi,\eta)$ can be extended further than α retaining the relation $G(\xi,\eta) = f^{-1}(F(\xi,\eta))$. Therefore, $\alpha = 1$, so $G(\xi,\eta)$ can be defined for all $\xi,\eta \in [0,1]$. Since F is differentiable and f^{-1} is locally differentiable, $G(\xi,\eta)$ is also differentiable.

Now, since $F(\xi,1) = y_0$ for all $\xi \in [0,1]$,

$$0 = \frac{d}{d\xi}F(\xi,1) = \frac{d}{d\xi}f(G(\xi,1)) = f'(G(\xi,1))\frac{d}{d\xi}G(\xi,1),$$

and, hence, the assumption implies that $\frac{d}{d\xi}G(\xi,1) = 0$, which implies $G(0,1) = G(1,1)$. Similarly, we have $\frac{d}{d\eta}G(0,\eta) = 0$, which implies $G(0,0) = G(0,1)$. Thus, $x_0 = G(0,0) = G(1,1) = x_1$. Therefore, f is a bijection. The continuity of f^{-1} follows from

$$\|f^{-1}(x_1) - f^{-1}(x_2)\| \leq \|x_1-x_2\| \cdot \sup_{x \in E} \|f'(x)^{-1}\|.$$

Therefore, by (3.2.6), $f^{-1} \in C^1(F,E)$.

To examine the condition 3° further, let us denote by $BC^1(E,F)$ the set of all $f \in C^1(E,F)$ such that $\sup_{x \in E} \|f'(x)\| < \infty$.

(3.5.7) $f \in BC^1(E,E)$ <u>is invertible if and only if it satisfies</u> 3°.

PROOF. Since the necessity is evident, assume that $f \in BC^1(E,E)$ satisfies 3°. Then by (3.5.6), f^{-1} exists in $C^1(E,E)$ and

$$\sup_{x \in E} \|(f^{-1})'(x)\| \leq \sup_{x \in E} \|f'(x)^{-1}\| < \infty,$$

which implies $f^{-1} \in BC^1(E,E)$.

NOTE: Theorems (3.5.1), (3.5.2) and (3.5.3) have been taken from Bartle [1], Section 21, and the following two theorems are from Lang [1]. See also Eells [1]. Theorem (3.5.6) was first proved by J.T. Schwartz [1], and then, with some generalizations, by Rheinboldt [1].

CHAPTER 4. DIFFERENTIABILITY OF SEMI-NORMS

Let $E \in LCS$ and $p \in P(E)$.

§4.1 Hadamard differentiability of semi-norms

All semi-norms, being convex functions, have differentiable properties to some degree. We start the whole discussion by the following theorem.

(4.1.1) For any $a,x \in E$, the right-hand derivative

$$p'_+(a)(x) = \lim_{\varepsilon \to 0+} \varepsilon^{-1}[p(a+\varepsilon x)-p(a)]$$

exists and has the following properties :

1°. $p'_+(a,a) = p(a)$ and $p'_+(a)(x) \leq p(x)$ for all $x \in E$;

2°. $p'_+ : E \to R$ is continuous, positively homogeneous and subadditive;

3°. the left-hand derivative $p'_-(a)(x)$ exists and $p'_-(a)(x) = - p'_+(a)(-x)$.

PROOF. The existence of the limit follows from the fact that $\{\xi^{-1}[p(a+\xi x)-p(a)]\}$ is monotone increasing with respect to $\xi > 0$. The remainder is easy to prove.

From this theorem it follows that, if $p \in D(a,R;\to E)$, then $p'(a) \in \bar{E}$, and, hence, $p \in D_G(a,R)$. Combining this with (1.4.6), we have the following theorem.

(4.1.2) If $p \in D(a,R;\to E)$, then $p \in D_H(a,R)$.

To give another criterion for Hadamard differentiability, we shall introduce some new terminology.

[4.1.3] Let $a \neq 0$. Then, $\bar{a} \in \bar{E}$ is called a supporting functional of p at a if

$$<a,\bar{a}> = p(a) \quad \text{and} \quad <x,\bar{a}> \leq p(x) \quad \text{for all} \quad x \in E.$$

(4.1.4) <u>Let</u> $a \neq 0$. <u>Then</u>, $p \in D_H(a,R)$ <u>if and only if there exists a unique supporting functional of</u> p <u>at</u> a.

PROOF. Let $\bar{a} \in \bar{E}$ be a supporting functional of p at a. Then,

$$p(a+\varepsilon x) - p(a) \geq \, <a+\varepsilon x, \bar{a}> \, - \, <a, \bar{a}> \, = \varepsilon <x, \bar{a}> \quad \text{if } x \in E,$$

which implies

$$p'_+(a)(x) \geq \, <x, \bar{a}> \, \geq p'_-(a)(x) \quad \text{for all } x \in E.$$

Therefore, if $p \in D_H(a,R)$, we have $p'(a) = \bar{a}$.

Conversely, if $p \notin D_H(a,R)$, there exists $b \in E$ such that $p'_-(a)(b) < p'_+(a)(b)$. Then, since a and b are linearly independent, there is a supporting functional $\bar{b} \in \bar{E}$ such that $<b, \bar{b}> = \alpha$, where α is any number between $p'_-(a)(b)$ and $p'_+(a)(b)$. Therefore, the supporting functional of p at a is not unique.

As the converse of (1.4.2), we have the following theorem.

(4.1.5) If $E \in BS$ and its norm p is Hadamard differentiable on $E \setminus \{0\}$, then $p'(x)(y)$ is jointly continuous with respect to $(x,y) \in (E \setminus \{0\}) \times E$.

PROOF. At first, we prove that $x_n \to a \neq 0$ and $b \in E$ imply $p'(x_n)(b) \to p'(a)(b)$. We can assume that $\|a\| = 1$.

Since $\|p'(x_n)\| = 1$ and the unit sphere of $\bar{E}$ is weak*-compact, there exists a weak* cluster point $\bar{x} \in \bar{E}$ of $\{p'(x_n)\}$ such that $\|\bar{x}\| = 1$. Then,

$$|<a,\bar{x}> - 1| \leq |<a,\bar{x}> - <a, p'(x_n)>| + |<x_n, p'(x_n)> - <a, p'(x_n)>|$$

$$\leq |<a,\bar{x}> - <a, p'(x_n)>| + p(x_n - a),$$

which implies, by (4.1.4), that $\bar{x} = p'(a)$.

Now, if $x_n \to a \neq 0$ and $y_n \to b$,

$$|p'(x_n)(y_n) - p'(a)(b)|$$

$$\leq |p'(x_n)(b) - p'(a)(b)| + |p'(x_n)(y_n) - p'(x_n)(b)|$$

$$\leq |p'(x_n)(b) - p'(a)(b)| + p(y_n-b) \to 0.$$

The following theorem is the fundamental result on Hadamard differentiability of norms of BS.

(4.1.6) (Mazur) *If a Banach space* E *is separable, its norm* p *is Hadamard differentiable in a dense* G_δ-*set*.

PROOF. Let $\{e_n\}$ be a dense subset and M be the set of points where p is Hadamard differentiable. Put.

$$M_n = \{x \in E : \lim_{\varepsilon \to 0} \varepsilon^{-1}[p(x+\varepsilon e_n) - p(x)] \text{ exists}\},$$

then $M \subset M_n$ for each n. If $a \in \bigcap_{n=1}^{\infty} M_n$, by (4.1.1),

$$p'_+(a)(e_n) = p'_-(a)(e_n) \quad \text{for all} \quad n.$$

Since $p'_+(a)$ is continuous,

$$p'_+(a)(x) = p'_-(a)(x) \quad \text{for all} \quad x \in E,$$

which means that $a \in M$. Hence, $M = \bigcap_{n=1}^{\infty} M_n$. Next, put

$$M_{n,i,j} = \{x \in E : \varepsilon^{-1}[p(x+\varepsilon e_n)-p(x)] - (-\varepsilon)^{-1}[p(x-\varepsilon e_n)-p(x)] \geq \frac{1}{i}$$

$$\text{if} \quad 0 < \varepsilon \leq 1/j\},$$

then, these sets are closed and $E \setminus M_n = \bigcup_{i,j=1}^{\infty} M_{n,i,j}$. Hence, M_n are G_δ and, consequently, M is G_δ.

To show that M is dense, we only need to prove that

$$E \setminus M = \bigcup^{\infty} (E \setminus M_n)$$

is of the first category. Now, if this is not the case, there is n for which $E \setminus M_n$ contains an interior point a, i.e., $U(a,\gamma) \subset E \setminus M_n$. Let us consider the function

$$\alpha(\xi) = p(a+\xi e_n) ,$$

which, by (4.1.1), has both one-side derivatives at each ξ. Therefore, it is differentiable except for countably many points. Hence, there exists ξ_0 such that $0 < |\xi_0| < \gamma p(e_n)^{-1}$ and α is differentiable at ξ_0. Then, $x_0 = a + \xi_0 e_n \in M_n$. On the other hand, $p(x_0-a) = |\xi_0 p(e_n)| < \gamma$, which means $x_0 \in U(a,\gamma)$, a contradiction.

The following theorem of Spatz [1] shows that even a simple type of differentiability of norms of Banach algebras gives a heavy restriction to their algebraic structures.

(4.1.7) <u>Let</u> E <u>be a real Banach algebra with the identity</u> e <u>and</u> $\|e\| = 1$. <u>If the limit</u>

$$\lim_{\varepsilon \to 0} \varepsilon^{-1}[\|e+\varepsilon x\|-\|e\|]$$

<u>exists for each singular element</u> x, <u>then</u> E <u>is a division algebra</u>.

PROOF. Let us put

$$p'_{\pm}(x)(y) = \lim_{\varepsilon \to 0\pm} \varepsilon^{-1}[\|x+\varepsilon y\|-\|x\|].$$

If x is singular, then

$$\|e+\varepsilon x\| \geq 1 \quad \text{for each} \quad \varepsilon \neq 0,$$

which implies $p'_+(e)(x) \geq 0$ and $p'_-(e)(x) \leq 0$. Therefore, $e + x$ is invertible, because if $e + x$ is singular, we have

$$0 = p'(e)(e+x) = p'(e)(e) + p'(e)(x) = p'(e)(e) = 1.$$

Consequently, for each $\alpha \in R$, $u_\alpha = \|e+\alpha x\|^{-1}(e+\alpha x)$ is invertible, and $\|u_\alpha\| = 1$. Now, since

$$\|u_\alpha + \xi x\| \leq \|e + \xi u_\alpha^{-1} x\|,$$

we have

$$p'_+(u_\alpha)(x) \leq p'_+(e)(u_\alpha^{-1} x) = 0 \quad \text{and} \quad p'_-(u_\alpha)(x) = p'_-(e)(u_\alpha^{-1} x) = 0,$$

which imply $p'(u_\alpha)(x) = 0$. Therefore

$$1 = p'(u_\alpha)(u_\alpha) = \|e+\alpha x\|^{-1} p'(u_\alpha)(e)$$

$$\leq \|e+\alpha x\|^{-1} \|u_\alpha\| \|e\| = \|e+\alpha x\|^{-1},$$

from which it follows that $1 \geq \|e+\alpha x\| \geq \|\alpha x\| - 1$. Since α is arbitrary we have $x = 0$.

Here we add a result which will be used later.

(4.1.8) <u>Let</u> $E \in NLS$, p <u>be its norm and</u> q <u>be the conjugate norm. If</u> $p(a) = 1$, $p \in D_H(a,R)$ <u>and</u> $q \in D_H(p'(a),R)$, <u>then</u>

$$q'(p'(a)) = i(a),$$

<u>where</u> i <u>is the natural embedding of</u> E <u>into</u> $\bar{\bar{E}}$.

PROOF. By (4.1.1), $q'(p'(a))$ and $i(a)$ are supporting functionals of q at $p'(a)$. Therefore, by (4.1.4), they have to be equal.

NOTE: (4.1.1) has been taken from Köthe [1, p.347]. The important fact (4.1.4) was first proved by Mazur [1], which also contains (4.1.6). In a sense, the following theorem of Day [1] generalizes (4.1.4). Theorem : <u>If a Banach space</u> E <u>is separable, there is an equivalent norm which is Hadamard differentiable at every non-zero point</u>.

Since semi-norms are convex functions, the materials of this and the next section are special cases of the results on the differentiability of convex functions.

The most important work on this problem is Asplund-Rockafeller [1].

§4.2 The Fréchet differentiability of semi-norms

We start with the following well-known theorem of Smulyan [1].

(4.2.1) Let $p(a) = 1$. Then, $p \in D(a,R)$ if and only if the following condition is satisfied :

(#) "if $< a,\bar{x}_\sigma > \to 1$ and $< x,\bar{x}_\sigma > \leq p(x)$ for all $x \in E$, then the net $\{\bar{x}_\sigma\}$ is convergent."

PROOF. Assume that $p \in D(a,R)$, $< a,\bar{x}_\sigma > \to 1$ and $< x,\bar{x}_\sigma > \leq p(x)$ for all $x \in E$. If $\bar{x}_\sigma \not\to p'(a)$, there is $\{x_\sigma\} \in B(E)$ such that $|< x_\sigma,\bar{x}_\sigma - p'(a) >| \not\to 0$. We can assume that $1 - < a,\bar{x}_\sigma >$ is always positive, because if $1 = < a,\bar{x}_\sigma >$ for some σ, then $\bar{x}_\sigma$ is a supporting functional of p at a, and then (4.1.4) implies $\bar{x} = p'(a)$. Now, put

$$\varepsilon_\sigma = (1-< a,\bar{x}_\sigma >)^{\frac{1}{2}} \text{ sign } < x_\sigma,\bar{x}_\sigma - p'(a) >.$$

Then, $\varepsilon_\sigma \to 0$, $\varepsilon_\sigma \neq 0$ and

$$r(p,a,\varepsilon_\sigma x_\sigma) = p(a+\varepsilon_\sigma x_\sigma) - p(a) - p'(a)(\varepsilon_\sigma x_\sigma)$$

$$\geq < a+\varepsilon_\sigma x_\sigma,\bar{x}_\sigma > - 1 - p'(a)(\varepsilon_\alpha x_\alpha)$$

$$= < a,\bar{x}_\sigma > - 1 + \varepsilon_\sigma < x_\sigma,\bar{x}_\sigma - p'(a) >,$$

which implies

$$|< x_\sigma,\bar{x}_\sigma - p'(a) >| \leq |\varepsilon_\sigma^{-1} r(p,a,\varepsilon_\sigma x_\sigma)| + |\varepsilon_\sigma^{-1}(1-< a,\bar{x}_\sigma >)| \to 0,$$

a contradiction.

Conversely, assume that (#) is satisfied. If $\bar{a}_1$ and $\bar{a}_2$ are supporting functionals of p at a, then the sequence $\{\bar{a}_1,\bar{a}_2,\bar{a}_1,\bar{a}_2,\ldots\}$ satisfies the assumption of (#). Hence, $\bar{a}_1 = \bar{a}_2$, and by (4.1.4), $p \in D_H(a,R)$. If $p \notin D(a,R)$, there are $\{\varepsilon_n\} \in (c_0)$, $\{x_n\} \in B(E)$ and $\delta > 0$ such that

$$|\varepsilon_n^{-1} r(p,a,\varepsilon_n x_n)| > \delta ,$$

where we can assume that $\varepsilon_n > 0$. By the Hahn-Banach theorem, we can take $\{\bar{x}_n\} \subset \bar{E}$ such that

$$p(a+\varepsilon_n x_n) = <a+\varepsilon_n x_n, \bar{x}_n> \text{ and } <x, \bar{x}_n> \leq p(x) \text{ for all } x \in E.$$

Then $\{\bar{x}_n\}$ satisfies the assumption of (#). Hence it converges, and, since the limit is also a supporting functional of p at a, it has to be equal to $p'(a)$. Therefore,

$$0 \leq \varepsilon_n^{-1} r(p,a,\varepsilon_n x_n) \leq \varepsilon_n^{-1}[<a+\varepsilon_n x_n, \bar{x}_n> - <a, \bar{x}_n> - <\varepsilon_n x_n, p'(a)>]$$

$$= <x_n, \bar{x}_n - p'(a)> \to 0,$$

which is a contradiction.

It is obvious from the proof that, in (#), net can be replaced by sequence.

Now, put $N_p = \{x \in E : p(x) = 0\}$.

(4.2.2) $p \in D(E \setminus N_p, R)$ <u>if and only if</u> $p \in D_H(E \setminus N_p, R)$ <u>and the mapping</u> $p' : E \setminus N_p \to \bar{E}$ <u>is continuous</u>.

PROOF. Let $p \in D(E \setminus N_p, R)$, $x_\sigma \in E \setminus N_p$ and $x_\sigma \to a \in E \setminus N_p$. Then, since $p(a) \neq 0$, we can assume that $p(a) = 1$. Then,

$$|<a, p'(x_\sigma)> - 1| = |<x_\sigma, p'(x_\sigma)> + <x_\sigma - a, p'(x_\sigma)> - 1|$$

$$\leq |p(x_\sigma) - 1| + p(x_\sigma - a) \to 0.$$

Hence, by (#), $p'(x_\sigma)$ converges to a limit, which is a supporting functional of p at a. Thus, $p'(x_\sigma) \to p'(a)$.

The converse follows from (1.4.4).

In particular, if p is differentiable, it is automatically C^1. In fact, we can go further.

(4.2.3) If a semi-norm p is C^1 at a, then it is EC^1 at a.

PROOF. We have to show that there is $U_a \in N(E)$ such that $p'(a+U_a)$ is a equicontinuous subset of $\bar{E}$, or, for any $\varepsilon > 0$ there is $U \in N(E)$ such that $|p'(a+U_a)(U)| < \varepsilon$. By (4.1.1) 1°, we need only to take $U = \{x : p(x) < \varepsilon\}$.

When $E \in BS$, the condition (#) is implied by

(##) if weak-limit $\bar{x}_n = \bar{x}_0$ and $\lim_{n\to\infty} \|\bar{x}_n\| = \|\bar{x}_0\|$, then $\lim_{n\to\infty} \bar{x}_n = \bar{x}_0$.

As is well-known, if $\bar{E}$ is uniformly convex, then (##) is satisfied. However, the following theorem of Klee [1] is essentially related to the condition (##). Theorem. If $E \in BS$ and $\bar{E}$ is separable, there is an equivalent norm which satisfies (##). Combining this theorem with (4.2.2), we arrive at the result of Restrepo.

(4.2.4) A separable Banach space has an equivalent C^1 norm at every non-zero point if and only if the conjugate space is separable.

PROOF. Let p be the C^1 norm of the Banach space E. Then, the separable set $\{p'(x) : x \in E \setminus \{0\}\}$ contains all support functionals of p at every point in the unit ball. By the Bishop-Phelps theorem (see NOTE below), it is dense. Hence, $\bar{E}$ itself is separable.

Conversely, let $\bar{E}$ be separable and let p be the norm obtained in the Klee's theorem cited above. Then, p satisfies (#) and, hence, it is C^1.

The following theorem shows that the Banach spaces with Fréceht differentiable norm are rather restricted.

(4.2.5) If $E \in BS$ and the norm of $\bar{E}$ is Fréchet differentiable on the surface of the unit sphere, then E is reflexive.

PROOF. Since E is complete, the condition (#) implies that, for $\bar{x} \in \bar{E}$ such that $\|\bar{x}\| = 1$, there exists $x \in E$ such that $<x,\bar{x}> = 1$ and $\|x\| = 1$, i.e., the unique supporting functional of the conjugate norm at each $\bar{x} \in \bar{E}$ with $\|\bar{x}\| = 1$ belongs to E, or, putting

$$M = \{\bar{\bar{x}} \in \bar{\bar{E}} : <\bar{x},\bar{\bar{x}}> = \|\bar{\bar{x}}\| \text{ for some } \bar{x} \text{ such that } \|\bar{x}\| = 1\},$$

we have $M \subset \bar{\bar{E}}$ as sets. On the other hand, the Bishop-Phelps theorem implies that M is dense in $\bar{\bar{E}}$. Since E is closed in $\bar{\bar{E}}$, we have $E = \bar{\bar{E}}$.

NOTE: (4.2.1) is a semi-norm version of the result of Smulyan [1]. (4.2.2) has been proved by Giles [1] and, in more general situation, by Asplund-Rockafellar [1] and Daneš-Kolomý [1].

The Bishop-Phelps theorem, which plays an essential rôle in the proof of (4.2.4) is as follows : Theorem. <u>For any</u> $E \in BS$, <u>the set of</u> $\bar{x} \in \bar{E}$ <u>which attains its norm on the surface of the unit ball of</u> E <u>is dense in</u> $\bar{E}$.

(4.2.5) is due to Giles [1], and, in more general cases, to Asplund-Rockafellar [1]. Giles [2] gives an example which shows that a BS with Fréchet differentiable norm is not necessarily reflexive. For further study, see Giles [3,4].

§4.3 Higher derivatives of semi-norms

Let $E \in LCS$ and $p \in P(E)$.

(4.3.1) <u>Assume that</u> $p \in D_H^2(a,R)$. <u>Then</u>,

1°. $p^{(2)}(a)(x,x) \geq 0$ <u>for all</u> $x \in E$;

2°. $p^{(2)}(a)(a) = 0$ <u>and</u> $<a,p^{(2)}(a)(x)> = 0$ <u>for all</u> $x \in E$;

3°. $|\alpha|p^{(2)}(\alpha a) = p^{(2)}(a)$.

PROOF. 1° Since

$$p(a+\delta x) - p(a) \leq p(a+\epsilon x+\delta x) - p(a+\epsilon x) \text{ for } \epsilon > 0 \text{ and } \delta > 0,$$

we have $p'(a)(x) \leq p'(a+\varepsilon x)(x)$, which implies $p^{(2)}(a)(x,x) \geq 0$.

2°. Since $p'((1+\varepsilon)a) = p'(a)$ for $\varepsilon > 0$,

$$p^{(2)}(a)(a) = \lim_{\varepsilon \to 0} \varepsilon^{-1}[p'((1+\varepsilon)a) - p'(a)] = 0.$$

The second relation follows from (1.8.2).

3°. This again follows immediately from the definition of the derivatives.

(4.3.2) <u>Let</u> $p \in C^n(a,R)$. <u>Then</u>

1°. $p^{(n)}(a)(a) = -(n-2)p^{(n-1)}(a)$ <u>for</u> $n \geq 2$;

2°. $p^{(n)}(a)(a^{n-1}) = 0$ <u>for</u> $n \geq 2$.

PROOF. 1°. The case $n = 2$ follows from (4.3.1). 2°. Assume that the relation holds for $k < n$. Then, for any $x \in E$,

$$p^{(k)}(a+\varepsilon x)(a) - p^{(k)}(a)(a)$$

$$= p^{(k)}(a+\varepsilon x)(a+\varepsilon x) - p^{(k)}(a)(a) - \varepsilon p^{(k)}(a+\varepsilon x)(x)$$

$$= -(k-2)p^{(k-1)}(a+\varepsilon x) + (k-2)p^{(k-1)}(a) - \varepsilon p^{(k)}(a+\varepsilon x)(x).$$

Therefore, for any $x \in E$,

$$p^{(k+1)}(a)(x)(a) = -(k-2)p^{(k)}(a)(x) - p^{(k)}(a)(x)$$

$$= -(k-1)p^{(k)}(a)(x).$$

Hence, the relation holds for n.

2°. Since this is true for $n = 2$, we have

$$p^{(n)}(a)(a^{n-1}) = -(n-2)p^{(n-1)}(a)(a^{n-2}) = \ldots$$

$$= (-1)^{n-2}(n-2)! \; p^{(2)}(a)(a) = 0.$$

Next, we prove a higher derivative version of (4.1.8).

(4.3.3) Let $E \in BS$, p be its norm and q be the conjugate norm. Assume that p and q are twice Hadamard differentiable on the unit spheres. Then, if $p(a) = 1$,

    1°. $\quad q''(p'(a))(p''(a)(x)) = x - p'(a)(x)a \quad$ for all $x \in E$;

    2°. $\quad p''(a)(q''(p'(a))(\bar{x})) = \bar{x} - <a,\bar{x}> p'(a) \quad$ for all $\bar{x} \in \bar{E}$.

PROOF. By (4.2.5), E is reflexive. Therefore, we may identify $\bar{\bar{E}}$ with E.

    1°. For any $x \in E$,

$$q'(p'(a+\varepsilon x)) - q'(p'(a)) = q''(p'(a))(p'(a+\varepsilon x) - p'(a))$$
$$+ r(q',p'(a),p'(a+\varepsilon x) - p'(a)).$$

On the other hand, by (4.1.8),

$$q'(p'(a+\varepsilon x)) = p(a+\varepsilon x)^{-1}(a+\varepsilon x) \quad \text{and} \quad q'(p'(a)) = a.$$

Thus, we have

$$q''(p'(a))(p''(a)(x)) = \lim_{\varepsilon \to 0} \varepsilon^{-1} \left| \frac{a+\varepsilon x}{p(a+\varepsilon x)} - a - r(q',p'(a),p'(a+\varepsilon x) - p'(a)) \right|$$
$$= x - p'(a)(x)a.$$

    2°. We have only to change the rôles of p and q.

Now we can prove a characterization of Hilbert spaces.

(4.3.4) Let $E \in BS$, p be its norm and q be the conjugate norm. If p and q are twice differentiable, E is isomorphic to a Hilbert space.

PROOF. Let $p(a) = 1$ and $E_a = \{x \in E : p'(a)(x) = 0\}$. Then, by (4.3.3). 1°,

$$\|p''(a)(x)\| \geq 1/\|q''(p'(a))\| \quad \text{if } x \in E_a.$$

Then, by (1.8.2) and (4.3.1). 1°,

$$\inf \{p''(a)(x,x) : x \in E_a \text{ and } \|x\| = 1\} > 0.$$

For any $x_i \in E$ (i=1,2), there are $y_i \in E_a$ and $\xi_i \in R$ uniquely such that $x_i = y_i + \xi_i a$. We define an inner product by

$$(x_1, x_2) = p''(a)(y_1, y_2) + \xi_1 \xi_2.$$

Then, the corresponding norm is equivalent to p.

NOTE: The results in this section are essentially due to Sundaresan [1]. Theorem (4.3.4) has also been proved by Bonic-Reis [1] and Rao [2].

§4.4 Differentiability of the supremum norms of function spaces

We start with the space of continuous functions. Let X be a compact Hausdorff space and $E \in BS$. The norm of E will be denoted by $p(x)$ or $\|x\|$ for $x \in E$.

We denote the set of all continuous functions of X into E by $C(X,E)$. For $f \in C(X,E)$, we can define a norm p_∞ by

$$p_\infty(f) = \sup_{x \in X} \|f(x)\|.$$

[4.4.1] A function $f : X \to E$ is said to be a peaking function if there is a point $a \in X$ such that $\|f(a)\| > \|f(x)\|$ when $x \neq a$. The point a is called the peak point of f.

(4.4.2) The norm p_∞ is Hadamard differentiable at f if and only if the following two conditions are satisfied :

1°. f is a peaking function with the peak point a.
2°. $p \in D_H(f(a), R)$.

Moreover, if $p_\infty \in D_H(f, R)$ and a is the peak point of f, then

$$p_\infty'(f)(g) = p'(f(a))(g(a)) \quad \text{for any } g \in C(X,E).$$

PROOF. Assume that $p_\infty \in D_H(f,R)$. Since X is compact, there is $a \in X$ such that $p_\infty(f) = \|f(a)\|$. Then, for any $g \in C(X,E)$,

$$p_\infty(f+\varepsilon g) - p_\infty(f) \geq \|f(a) + \varepsilon g(a)\| - \|f(a)\|,$$

which implies

$$p'_\infty(f)(g) \geq p'_+(f(a))(g(a)) \quad \text{and} \quad p'_\infty(f)(g) \leq p'_-(f(a))(g(a)).$$

Therefore, by (4.1.2), $p \in D_H(f(a),R)$ and

$$p'_\infty(f)(g) = p'(f(a))(g(a)).$$

To show that a is unique, assume that $p_\infty(f) = \|f(b)\|$ and $a \neq b$. Then, $p'(f(a))(g(a)) = p'(f(b))(g(b))$ for any $g \in C(X,E)$. We arrive at a contradiction if we take g such that $g(a) = f(a)$ and $g(b) = 0$.

Conversely, assume that 1° and 2° are satisfied. Let $g \in C(X,E)$ and $\varepsilon_n \to 0$. Then, there exists $x_n \in X$ such that

$$p_\infty(f+\varepsilon_n g) = \|f(x_n) + \varepsilon_n g(x_n)\|.$$

It is easy to see that $x_n \to a$. Now, since

$$p_\infty(f+\varepsilon_n g) - p_\infty(f) \geq \|f(a) + \varepsilon_n g(a)\| - \|f(a)\|$$

$$= \|f(a) + \varepsilon_n g(a)\| - p'(f(a))(f(a)) \geq \varepsilon_n p'(f(a))(g(a)),$$

which implies $(p_\infty)'_+(f)(g) \geq p'(f(a))(g(a))$.

On the other hand,

$$p_\infty(f+\varepsilon_n g) - p_\infty(f) \leq \|f(x_n) + \varepsilon_n g(x_n)\| - \|f(x_n)\|$$

$$= \varepsilon_n p'(f(x_n) + \theta_n \varepsilon_n g(x_n))(g(x_n))$$

for some $\theta_n \in (0,1)$. By (4.1.5), we have $(p_\infty)'_+(f)(g) \leq p'(f(a))(g(a))$. Hence, $(p_\infty)'_+(f)(g) = p'(f(a))(g(a))$. We have the same result for $(p_\infty)'_-(f)(g)$. Therefore, $p_\infty \in D_H(f,R)$.

For Fréchet differentiability, we have the following theorem.

(4.4.3) $p_\infty \in D(f,R)$ <u>if and only if the following three conditions are satisfied</u>:

1°. $\quad$ f <u>is a peaking function</u>;
2°. $\quad$ <u>the peak point</u> a <u>is an isolated point</u>;
3°. $\quad$ $p \in D(f(a),R)$.

PROOF. $\quad$ Assume that $p_\infty \in D(f,R)$. Then, by the previous theorem, $p \in D_H(f(a),R)$ and there is a peak point a of f such that

$$p'_\infty(f)(g) = p'(f(a))(g(a)) \quad \text{for all} \quad g \in C(X,E).$$

Now, assume that a is not isolated. Then, $x_n \to a$ for some $x_n \in X$. We can assume $f(x_n) \neq 0$ and $x_n \neq a$. Take continuous functions $\alpha_n : X \to [0,1]$ such that $\alpha_n(x_n) = 1$ and $\alpha_n(x) = 0$ in a neighbourhood (depending on n) of a, and put

$$g_n(x) = 2\|f(x_n)\|^{-1}\alpha_n(x)\|f(x_n) - f(a)\|f(x_n).$$

Then, $p_\infty(g_n) \to 0$. However, since $g_n(a) = 0$, we have

$$p'_\infty(f)(g_n) = p'(f(a))(g_n(a)) = 0.$$

This is impossible, because

$$p_\infty(g_n)^{-1}[p_\infty(f+g_n) - p_\infty(f)] \geq \frac{1}{2}.$$

Thus, a is isolated.

Next, we shall prove 3°. Since a is isolated, we can define $e : E \to C(X,E)$ such that

$$e(z)(a) = z \quad \text{and} \quad e(z)(x) = 0 \quad \text{if} \quad x \neq a.$$

Then, e is continuous and linear, and, moreover, $p_\infty \circ e = p$. Therefore, if p_∞ is differentiable at $e(f(a))$, we have $p \in D(f(a),R)$. To show that $p_\infty \in D(e(f(a)),R)$, we take $\delta > 0$ such that $\delta < \|f(a)\|/2$. Then, if $p_\infty(g) < \delta$,

$$\|f(a) + g(a)\| - p_\infty(g) \geq \|f(a)\| - \|g(a)\| - p_\infty(g) \geq \|f(a)\| - 2p_\infty(g) > 0,$$

or,

$$\|g(x)\| \leq \|f(a) + g(a)\| \quad \text{for all } x \in X.$$

Therefore, $p_\infty(e(f(a))+g) = \|f(a) + g(a)\|$. Moreover, since a is isolated if we take δ small enough, we have

$$p_\infty(f+g) = \|f(a) + g(a)\| \quad \text{if } p_\infty(g) < \delta.$$

Thus, if $p_\infty(g) < \delta$, we have $p_\infty(e(f(a))+g) = p_\infty(f+g)$, which implies $p_\infty \in D(e(f(a)),R)$.

Conversely, assume that the conditions 1°, 2° and 3° are satisfied.

Then, by the previous theorem, $p_\infty \in D_H(f,R)$ and $p'_\infty(f)(g) = p'(f(a))(g(a))$ for every $g \in C(X,E)$. Since a is isolated and $p \in D(f(a),R)$, for any $\varepsilon > 0$ there exists $\delta > 0$ such that $p_\infty(g) < \delta$ implies

$$p_\infty(f+g) = \|f(a)+g(a)\|$$

and

$$\left| \|g(a)\|^{-1} [\|f(a)+g(a)\| - \|f(a)\| - p'(f(a))(g(a))] \right| < \varepsilon.$$

Therefore,

$$p_\infty(g)^{-1} \left| p_\infty(f+g) - p_\infty(f) - p'_\infty(f)(g) \right| < \varepsilon,$$

which shows that $p_\infty \in D(f,R)$.

As an immediate consequence, we have the following theorem.

(4.4.4) <u>The supremum norm of</u> $C([0,1],E)$ <u>is nowhere Fréchet differentiable</u>.

There may be a norm on $C([0,1],R)$ which is Fréchet differentiable at some points. However, since it is separable and the conjugate space is not separable, (4.2.2) and (4.2.5) imply the following theorem.

(4.4.5) <u>It is impossible to define a norm in</u> $C([0,1],R)$ <u>which is equivalent to the supremum norm and Fréchet differentiable at every non-zero point</u>.

Next, we investigate the space (c_0) of the sequence $x = \{x(n)\}$ of real numbers such that $\lim_{n \to \infty} x(n) = 0$. The supremum norm is defined by

$$p_\infty(x) = \sup_{n \geq 1} |x(n)|.$$

The terms "peaking sequence" and "peak point" are defined as in (4.4.1).

(4.4.6) <u>If the norm</u> p_∞ <u>is Hadamard differentiable at</u> $a \in (c_0)$, <u>then</u> a <u>is a peaking sequence and, for its peak point</u> n_0,

$$p'_\infty(a)(x) = x(n_0) \text{ sign } (a(n_0)).$$

<u>If</u> a <u>is a peaking sequence in</u> (c_0), p_∞ <u>is Fréchet differentiable there</u>.

PROOF. Assume that $p_\infty \in D_H(a,R)$. There always exists n_0 such that $p_\infty(a) = a(n_0)$. Then, the sequence $\bar{a} = \{\bar{a}(n)\}$, defined by

$$\bar{a}(n) = \text{sign } (a(n_0)) \text{ if } n = n_0; \; = 0 \text{ if } n \neq n_0,$$

belongs to $\ell_1 = \overline{(c_0)}$, and

$$< a, \bar{a} > = p_\infty(a) \text{ and } \|\bar{a}\| = 1 \text{ in } \ell_1,$$

which means that $\bar{a}$ is a supporting functional of p_∞ at a. Therefore, by (4.1.4), $p'_\infty(a) = \bar{a}$. Thus, a is a peaking sequence and we have the required equality.

Next, assume that a is a peaking sequence and n_0 is the peak point. Thus, $\bar{a}$ defined above is the unique supporting functional. There exists $\delta > 0$ such that $p_\infty(x) < \delta$ implies

$$p_\infty(a+x) = |a(n_0) + x(n_0)|.$$

Hence,

$$p_\infty(a+x) - p_\infty(a) - <x,\bar{a}> = 0,$$

which shows that $p_\infty \in D(a,R)$.

Since (c_0) and its conjugate space are separable, it follows from (4.1.4) that (c_0) has an equivalent C^1-norm. However, we have a stronger result.

(4.4.7) (c_0) <u>has a C^∞-norm which is equivalent to the supremum norm</u>.

PROOF. Put

$$\phi_1(\xi) = \exp(-\xi^2) \text{ for } \xi > 0; \ = 0 \text{ for } \xi \leq 0;$$

$$\phi_2(\xi) = \phi_1(\xi-1)\phi_1(2-\xi);$$

$$\phi_3(\xi) = \int_0^\xi \phi_2(\eta)d\eta \ \Big| \ \int_0^2 \phi_2(\eta)d\eta$$

and

$$\phi(\xi) = 1 - \phi_3(|\xi|).$$

For $x \in (c_0)$, put

$$\psi(x) = \prod_{n=1}^\infty \phi(x(n)).$$

Then, the Minkowski functional of the convex set

$$\{x \in (c_0) : \psi(x) \geq \frac{1}{2}\}$$

is the required norm.

Let (M) be the Banach space of all essentially bounded real-valued functions defined on $[0,1]$ with the following norm :

$$p_\infty(x) = \underset{0 \leq \xi \leq 1}{\text{ess. sup}} |x(\xi)|.$$

This space is not separable, and the following theorem shows that the separability assumption of (4.1.6) can not be dropped.

(4.4.8) The norm p_∞ of (M) is not Hadamard differentiable at any point.

PROOF. Assume that $p_\infty \in D_H(a,R)$. We can assume $p_\infty(a) = 1$. We shall show that we may assume that $a(\xi) \geq 0$ for all $\xi \in [0,1]$. To show this, what we have to show is $p_\infty \in D_H(a,R)$ implies $p_\infty \in D_H(|a|,R)$.

Now, let $\bar{a}$ be a supporting functional of p_∞ at $|a|$, and let χ_+ and χ_- be the characteristic functions of $\{\xi : a(\xi) > 0\}$ and $\{\xi : a(\xi) \leq 0\}$ respectively. Then,

$$\chi_+(\xi) + \chi_-(\xi) = 1 \text{ and } \chi_+(\xi)\chi_-(\xi) = \chi_-(\xi)\chi_+(\xi) = 0 \text{ for } \xi \in [0,1].$$

Let $T_+, T_- : (M) \to (M)$ be defined by

$$(T_\pm)(\xi) = \chi_\pm(\xi) x(\xi) \text{ for } \xi \in [0,1].$$

Then, $T_\pm$ are continuous and linear, and

$$|a| = T_+ a - T_- a, \quad a = T_+ a + T_- a.$$

Hence,

$$< a, \bar{a}(T_+ + T_-) > = < |a|, \bar{a} > = p_\infty(a),$$

from which it follows that $\|\bar{a}(T_+ + T_-)\| \geq 1$ in $(\tilde{M})$. On the other hand,

$$\|\bar{a}(T_+ + T_-)\| = \sup_{p_\infty(z)=1} |< T_+ z + T_- z, \bar{a} >| \leq \|\bar{a}\|,$$

because $p_\infty(T_+ z + T_- z) = 1$ if $p_\infty(z) = 1$. Therefore, $\bar{a}(T_+ + T_-)$ is a supporting functional of p_∞ at a.

Now, if there are two supporting functionals $\bar{a}$ and $\bar{b}$ of p_∞ at $|a|$, then $\bar{a}(T_+ + T_-)$ and $\bar{b}(T_+ + T_-)$ are supporting functionals of p_∞ at a. Hence, by the assumption

$$\bar{a}(T_+ + T_-) = \bar{b}(T_+ + T_-).$$

By applying $T_\pm$ from the right, we have

$$\bar{a}T_+ = \bar{b}T_+ \quad \text{and} \quad \bar{a}T_- = \bar{b}T_-,$$

which imply $\bar{a} = \bar{b}$. Hence, $p_\infty \in D_H(|a|, R)$.

Thus, we can assume that

$$\underset{0 \leq \xi \leq 1}{\text{ess. sup}} \, a(\xi) = 1$$

We take a sequence $\{I_n\}$ of measurable sets such that

$$|I_n| > 0, \quad I_n \cap I_m = \phi \quad \text{if} \quad m \neq n \quad \text{and} \quad a(\xi) \geq n/(n+1) \quad \text{if} \quad \xi \in I_n.$$

Put

$$\alpha_n(x) = |I_n|^{-1} \int_{I_n} x(\xi) d\xi \quad \text{for} \quad x \in (M).$$

Then, $\alpha_n \in \overline{(M)}$ and $\|\alpha_n\| = 1$ in $\overline{(M)}$. Put

$$A_1 = \{x \in (M) : \beta_1(x) = \lim_{n \to \infty} \alpha_{2n-1}(x) \text{ exists}\}$$

and

$$A_2 = \{x \in (M) : \beta_2(x) = \lim_{n \to \infty} \alpha_{2n}(x) \text{ exists}\}$$

Then, A_1 and A_2 are subspaces of $\overline{(M)}$, and β_i are continuous linear functionals on A_i ($i=1,2$) with $\|\beta_i\| = 1$. Moreover, $\beta_1 \neq \beta_2$, because, for the characteristic function of I_1, we have $\beta_1(x) \neq 0$ and $\beta_2(x) = 0$.

Let $\bar{b}_i$ be the extension of β_i over E. Then, $\|\bar{b}_i\| = 1$ and

$$< a, \bar{b}_1 > = \lim_{n \to \infty} \alpha_{2n-1}(a) = \lim_{n \to \infty} |I_{2n-1}| \int_{I_{2n-1}} a(\xi) d\xi \geq$$

$$\geq \lim_{n \to \infty} (2n-1)/2n = 1.$$

On the other hand, $< a, \bar{b}_1 > \leq \|a\| \|\bar{b}_1\| \leq 1$. Thus, $< a, \bar{b}_1 > = 1$. Therefore, $\bar{b}_1$ is a supporting functional of p_∞ at a. Since the same argument applies to $\bar{b}_2$, there are two different supporting functionals of p_∞ at a. Hence,

$p_\infty \notin D_H(a,R)$.

For the space (m) of all bounded sequences of real numbers with the supremum $p_\infty(x) = \sup_{n \geq 1} |x(n)|$, we can prove the following theorem in a similar manner to (4.4.6).

(4.4.9) The supremum norm p_∞ of (m) is Hadamard differentiable at $a \in (m)$ if and only if a is a peaking sequence. If this is the case, for the peak point n_0,

$$p'_\infty(a)(x) = x(n_0) \operatorname{sign}(a(n_0)).$$

NOTE: [4.4.1] is due to Cox-Nadler [1], which contains (4.4.3) and the related results. (4.4.7) was given by N.H. Kuiper and was presented in Banic-Frampton [1]. (4.4.8) is due to Mazur [1].

§4.5 Differentiability of norms of L^p-spaces

In $L^p[0,1] = L^p$ and ℓ^p, we define their norms as usual :

$$\|x\| = \left(\int_0^1 |x(\xi)|^p d\xi\right)^{1/p} \quad \text{and} \quad \|x\| = \left(\sum_{n=1}^\infty |x(n)|^p\right)^{1/p}.$$

Their "modulars" are defined by

$$m(x) = \|x\|^p \quad \text{in both cases.}$$

It follows from this relation that the norms are differentiable at a point if and only if their modulars are differentiable there, and, if this is the case, we have $m'(x) = pn'(x)$, where $n(x) = \|x\|$ for all $x \in E$.

(4.5.1) The norms of L^p and ℓ^p ($p>1$) are

1°. C^∞ at every non-zero point when p is an even integer;

2°. C^{p-1} at every non-zero point when p is an odd integer;

3°. $C^{[p]}$ at every non-zero point when p is not an integer, where $[p]$ is the largest integer smaller than p.

PROOF. Let n be the greatest integer such that $n < p$. Put $\phi(\xi) = |\xi|^p$. Then, $\phi \in C^n(R,R)$ and

$$\phi^{(k)}(\xi) = p(p-1)\ldots(p-k+1)|\xi|^{p-k}(\text{sign }(\xi))^k.$$

Therefore, by the Taylor's formula,

$$\phi(\xi+\epsilon) - \sum_{k=0}^{n} \phi^{(k)}(\xi)\epsilon^k/k!$$

$$= \int_0^1 [(1-\eta)^{n-1}(\phi^{(n)}(\xi+\eta\epsilon) - \phi^{(n)}(\xi))\epsilon^n/(n-1)!]d\eta.$$

Hence,

$$\left|\phi(\xi+\epsilon) - \sum_{k=0}^{n} \phi^{(k)}(\xi)\epsilon^k/k!\right|$$

$$\leq p(p-1)\ldots(p-n+1)|\epsilon|^p \int_0^1 [(1-\eta)^{n-1}\eta^{p-n}/(n-1)!]d\eta$$

$$\leq p(p-1)\ldots(p-n+1)|\epsilon|^p/n! .$$

Now, let $m(a) = 1$ and define a k-linear mapping $u_k(a)$ on L_p by

$$u_k(a)(x_1,\ldots,x_k) = \int_0^1 \phi^{(k)}(a(\xi))x_1(\xi)\ldots x_k(\xi)d\xi.$$

Then, since

$$|u_k(a)(x_1,\ldots,x_k)| \leq m(a)^{1-\frac{k}{p}} \|x_1\|\ldots\|x_k\|,$$

$u_k(a)$ is continuous. Moreover

$$\left|m(a+x) - m(a) - \sum_{k=1}^{n} u_k(a)(x^k)/k!\right|$$

$$\leq \int_0^1 \left|\phi(a(\xi)+x(\xi)) - \phi(a(\xi)) - \sum_{k=1}^{n} \phi^{(k)}(a(\xi))(x(\xi))^k\right|d\xi$$

$$\leq (p(p-1)\ldots(p-n+1)/n!)\int_0^1 |x(\xi)|^n|x(\xi)|^{p-n}d\xi$$

$$\leq (p(p-1)\ldots(p-n+1)/n!)m(x).$$

Therefore, by (1.8.5), m is C^n and $m^{(k)}(a)(x) = u_k(a)(x)$ for $k = 0,1,2,\ldots,n$.

If p is an even integer, then, since $n = p - 1$, $\phi^{(n)}(\xi) = p!.\xi$. Hence,

$$m^{(n)}(a)(x_1,\ldots,x_n) = p! \int_0^1 a(\xi)x_1(\xi)\ldots x_n(\xi)d\xi,$$

which shows that $m^{(n)}(a)$ is linear with respect to a. Therefore, $m^{(n+1)}$ is constant and $m^{(n+2)}(x) = 0$ for all $x \neq 0$. Thus, m is C^∞.

The case when $p = 1$ is exceptional. At first, we know, by (4.2.5), that L_1 and ℓ_1 do not have equivalent norms which are Fréchet differentiable at every non-zero point. On the other hand, by (4.1.6), their original norms are Hadamard differentiable in a dense subset and by Day's theorem in the Note of §4.1, they have equivalent norms which are Hadamard differentiable at every non-zero point.

(4.5.2) <u>The original norms of</u> L_1 <u>and</u> ℓ_1 <u>are Hadamard differentiable at</u> a <u>if and only if</u> $a(\xi) \neq 0$ <u>almost everywhere or</u> $a(n) \neq 0$ <u>for all</u> n <u>respectively</u>. <u>If this is the case</u>,

$$m'(a)(x) = \int_0^1 x(\xi) \text{ sign } (a(\xi))d\xi \quad \text{or} \quad = \sum_{n=1}^\infty x(n) \text{ sign } (a(n)).$$

PROOF. We prove the case of ℓ_1. If $m \in D_H(a,R)$ and $a(n_0) = 0$ for some n_0, then the element $\bar{a} \in \bar{\ell}_1$, defined by

$$\bar{a}(n) = \text{sign } (a(n)) \text{ if } a(n) \neq 0; \ = \alpha \text{ if } a(n) = 0,$$

is a supporting functional for any $\alpha \in R$, which is a contradiction. If $a(n) \neq 0$ for all n, then, for sufficiently small ε,

$$|a(n) + \varepsilon x(n)| - |a(n)| - \varepsilon \bar{a}(n)x(n) = 0 \quad \text{for all} \quad n,$$

where $\bar{a}(n) = \text{sign } (a(n))$ for all n. Thus, $m \in D_H(a,R)$.

NOTE: (4.5.1) has been proved by Banic-Frampton [2] and Sundaresan [1]. Both proofs rely on the work of Kurzweil [1]. These spaces are special cases of the Orlicz and modular spaces. Rao [1,3] have some results for Orlicz spaces.

CHAPTER 5. SMOOTHNESS

Roughly speaking, a topological linear space will be said to be smooth if it has a sufficiently big family of smooth functions. Corresponding to various degrees of differentiability, there are various kinds of smoothness. A method which enables us to treat these in a unified manner has been proposed by Bonic-Frampton [2] for the Banach spaces, and Lloyd [2,3] has dealt with its generalization to TLS.

In the following, we shall present their results for the case of LCS.

§5.1 S-categories

[5.1.1] A category S is called an S-category if it satisfies the following conditions :

1°. its class of objects, which we denote by O, consists of all open subsets of all LCSs;

2°. for $X,Y \in O$, the set $S(X,Y)$ of morphisms of X into Y satisfies the following conditions :

(S1) $EC^{\infty}(X,Y) \subset S(X,Y)$;

(S2) $f \in S(X,Y)$ and $f(X) \subset Z \in O$ imply $f \in S(X,Z)$;

(S3)   if, for any $x \in X$ there exists $Z \in O$ such that $x \in Z \subset X$ and $f/Z \in S(Z,Y)$, then $f \in S(X,Y)$;

(S4)   if $f_i \in S(X_i, Y_i)$ (i=1,2), then $f_1 \times f_2 \in S(X_1 \times X_2, Y_1 \times Y_2)$.

An S-category S is said to be continuous if

$$S(X,Y) \subset C(X,Y) \quad \text{for all} \quad X,Y \in O.$$

When we need to specify the space in which X is situated, we denote it by E_X. If $f \in S(X,Y)$, f is called an S-smooth mapping of X into Y.

Obviously, the category C that corresponds to $X,Y \in O$ the set $C(X,Y)$ of all continuous mappings of X into Y is an S-category. However, our main interest lies in the categories whose morphisms are differentiable mappings.

(5.1.2) D_H^n (n=1,2,...) are S-categories.

PROOF. The composition properties were proved in (1.2.9) and (1.8.3).
(S1) is a part of the definition of equicontinuous differentiability.
(S2) and (S3) are obvious. (S4) follows from (1.2.11).

(5.1.3) D^n (n=1,2,...) are S-categories.

PROOF. The same as (5.1.2).

(5.1.4) EC^n (n=1,2,...) are continuous S-categories.

PROOF. (1.9.4) and (1.9.5).

In general, C^n is not an S-category because the chain rule does not hold. By (1.9.2), if the spaces are normed, it is a continuous S-category. Further, D_H^n and D^n are continuous S-categories if the spaces are normed.

The following two properties will be used frequently.

(5.1.5) If $f \in S$ and $g \in EC^\infty$, then $f \circ g$ and $g \circ f$ belong to S whenever they are defined.

PROOF. Immediate from the definition of a category and (S1).

(5.1.6) If $f_i \in S(X,Y)$ (i=1,2,...,n) and $g \in EC^\infty(Y \times Y \times ... \times Y, Y)$, then, for the diagonal mapping $\Delta_n : X \to X \times ... \times X$ (n-times),

$$g \circ (f_1 \times ... \times f_n) \circ \Delta_n \in S(X,Y).$$

PROOF. Immediate from (5.1.5).

As an application of (5.1.6), we have the following fact.

(5.1.7) If Y is linear, the set $S(X,Y)$ is linear.

PROOF. Define $g : Y \times Y \to Y$ by $g(y_1, y_2) = \alpha y_1 + \beta y_2$. Then,

$$g \in L(Y \times Y, Y) \subset EC^\infty(Y \times Y, Y)$$

Thus, for $f_1, f_2 \in S(X,Y)$, by (5.1.6), $g \circ (f_1 \times f_2) \circ \Delta_2 \in S(X,Y)$.

NOTE: [5.1.1] is a LCS-version of the original definition of Bonic-Frampton [2], where the case of Banach spaces has been considered. As we have seen in §1.7 and §1.8, if the spaces are not normed, the set C^∞, which should be the best example of the S-category, is not necessarily a semigroup and C^∞-mappings are not always continuous. Therefore, we have replaced C^∞ by EC^∞ and dropped the assumption that every morphism is continuous, so that D_H^n, D^n and EC^n can be covered by the general theory of the S-category. Needless to say, when the spaces are normed, the definition [5.1.1] coincides with the original one of Bonic-Frampton [2].

Lloyd [3] has given a slightly different definition of the S-category, which also works effectively. He called a mapping $f \in C^1$ at a if $f \in C^1(a,F)$ in our definition and, moreover, there is $U_a \in N(E)$ such that $f'(a+U_a)$ is bounded in $L(E,F)$, and (S1) of his definition has C^∞ in this sense instead of EC^∞. Independently, Penot [1] has considered these two substitutes - EC^∞ and Lloyd's C^∞.

§5.2 S-smooth spaces

Let S be an S-category. We shall consider $E \in LCS$ for which $S(E,R)$ is sufficiently large. The exact meaning of its being sufficiently large is defined below. Let us recall that the weak topology defined by the family $S(E,R)$, which we shall call the $S(E,R)$-topology, is the weakest topology on E by which all members of $S(E,R)$ become continuous. A subset $A \subset E$ is $S(E,R)$-open if and only if, for any $a \in A$, there exist $f_i \in S(E,R)$ ($i=1,2,\ldots,n$) and $\epsilon > 0$ such that

$$\{x \in E : |f_i(x) - f_i(a)| < \epsilon \text{ for } i=1,2,\ldots,n\} \subset A.$$

Obviously, this topology need not be a linear topology.

[5.2.1] $E \in$ LCS is said to be S-smooth if its topology is weaker than the $S(E,R)$-topology.

If S is a continuous S-category, since all members of $S(E,R)$ are continuous, the $S(E,R)$-topology is not stronger than the original topology of E. Therefore,

(5.2.2) *If S is a continuous S-category, then E is S-smooth if and only if its topology is equivalent to the* $S(E,R)$-*topology.*

A convenient criterion for E to be S-smooth is provided by the following theorem.

(5.2.3) *E is S-smooth if and only if, for any* $A \in \mathcal{O}(E)$ *and* $a \in A$, *there exists* $f \in S(E,R)$ *such that* $0 \leq f \leq 1$ *and* $a \in \{x \in E : f(x) > 0\} \subset A$.

PROOF. Let E be S-smooth, $A \in \mathcal{O}(E)$ and $a \in A$. Then, since A is $S(E,R)$-open, there are $f_i \in S(E,R)$ and $\varepsilon > 0$ such that

$$\{x \in E : |f_i(x) - f_i(a)| < \varepsilon \ (1 \leq i \leq n)\} \subset A,$$

which means, for $\alpha_i = f_i(a) - \varepsilon$ and $\beta_i = f_i(a) + \varepsilon$,

$$\alpha_i < f_i(x) < \beta_i \quad (1 \leq i \leq n) \quad \text{imply} \quad x \in A.$$

Take $\phi \in C^\infty(R^n, R)$ such that

$$\phi(\xi_1, \ldots, \xi_n) > 0 \text{ if } \alpha_i < \xi_i < \beta_i \ (1 \leq i \leq n); = 0 \text{ otherwise,}$$

and consider

$$f = \phi \circ (f_1 \times \ldots \times f_n) \circ \Delta_n \in S(E,R).$$

Then, f satisfies the required condition.

The converse is obvious, because the set $\{x \in E : f(x) > 0\}$ is $S(E,R)$-open for any $f \in S(E,R)$.

(5.2.4) <u>Let</u> E,F ∈ LCS. <u>Assume that</u> F <u>is</u> S-<u>smooth and</u> S(E,F) <u>contains an</u> <u>open injection</u>. <u>Then</u>, E <u>is</u> S-<u>smooth</u>.

PROOF. Let h ∈ S(E,F) be an open injection and a ∈ A ∈ O(E). Since h(a) ∈ h(A) ∈ O(F), by (5.2.3), there is f ∈ S(F,R) such that h(a) ∈ {y ∈ F : f(y) > 0} ⊂ h(A) and $0 \leq f \leq 1$. Then,

$$f \circ h \in S(E,R) \text{ and } a \in \{x \in E : (f \circ h)(x) \neq 0\} \subset A,$$

which means that E is S-smooth.

This fact reminds us of the following notion given by Pelcynski [1].

[5.2.5] E,F ∈ BS are said to be <u>almost</u> C^n-<u>diffeomorphic</u> if there are homeomorphisms f : E → F and g : F → E such that f and g are C^n.

The reason for the existence of this notion, which preserves the C^n-smoothness, is clear from the following theorem.

(5.2.6) 1°. <u>There are two Banach spaces which are not isomorphic but are almost</u> C^n-<u>diffeomorphic for all</u> n.

2°. <u>There are topologically homeomorphic Banach spaces which are not almost</u> C^1-<u>diffeomorphic</u>.

For the proof, see Pelcynski [1].

The following theorem is a convenient tool for the study of smoothness.

(5.2.7) <u>Let</u> E <u>have the projective topology determined by</u> $\{E_\sigma, \pi_\sigma\}$ (see §1.5). <u>If each</u> E <u>is</u> S-<u>smooth, then</u> E <u>is</u> S-<u>smooth</u>. <u>The converse is not true</u>.

PROOF. Let a ∈ A ∈ O(E). There are $U_i \in N(E_{\sigma_i})$ ($1 \leq i \leq n$) such that

$$a + \bigcap_{i=1}^{n} \pi_{\sigma_i}^{-1}(U_i) \subset A.$$

By the assumption, there are $f_i \in S(E_{\sigma_i}, R)$ such that

$$\pi_{\sigma_i}(a) \in \{x \in E_{\sigma_i} : f_i(x) > 0\} \subset \pi_{\sigma_i}(a) + U_i \quad (1 \leq i \leq n).$$

Take α_i and β_i such that

$$0 < \alpha_i < f_i(\pi_{\sigma_i}(a)) < \beta_i \quad (1 \leq i \leq n),$$

and take $\phi \in C^\infty(R^n, R)$ such that

$$\phi(\xi_1, \ldots, \xi_n) > 0 \text{ if } \alpha_i < \xi_i < \beta_i \ (1 \leq i \leq n); \ = 0 \text{ otherwise.}$$

Define $f : E \to R$ by

$$f = \phi \circ (f_1 \times \ldots \times f_n) \circ (\pi_{\sigma_1} \times \ldots \times \pi_{\sigma_n}) \circ \Delta_n.$$

Then, f satisfies the conditions of (5.2.3).

To see that the converse is not true, consider the space ℓ^∞ and its subspace (c_0). As we shall see in (5.2.9), (c_0) is C^∞-smooth, but ℓ^∞ is not.

We have another convenient criterion for smoothness via differentiability of semi-norms, namely (5.2.9). We need some new terminology.

[5.2.8] E is said to be **strongly** S-**smooth** if there exists a family $P_S(E)$ of continuous semi-norms which generates the topology of E and

$$p \in S(E \setminus N_p, R) \text{ for each } p \in P_S(E),$$

where $N_p = \{x \in E : p(x) = 0\}$.

(5.2.9) **If** E **is strongly** S-**smooth, then** E **is** S-**smooth**.

PROOF. Let $a \in A \in \mathcal{O}(E)$. There are $p_i \in P_S(E)$ $(1 \leq i \leq n)$ and $\varepsilon > 0$ such that

$$a + \{x \in E : \sup_{1 \leq i \leq n} p_i(x) < \varepsilon\} \subset A.$$

Take $\phi \in C^\infty(R, R)$ such that $\phi(\xi) = 1$ if $|\xi| < \varepsilon/2$ and $\phi(\xi) = 0$ if $|\xi| \geq \varepsilon$. Let α and β be such that $0 < \alpha < 1 < \beta$ and choose $\phi \in C^\infty(R^n, R)$ such that

$0 \leq \psi \leq 1$ and

$$\psi(\xi_1,\ldots,\xi_n) > 0 \text{ if } \alpha < \xi_i < \beta \text{ } (1 \leq i \leq n); = 0 \text{ otherwise.}$$

Then, the mapping $f : E \to R$ defined by

$$f = \psi \circ ((\phi \circ p_1) \times \ldots \times (\phi \circ p_n)) \circ \Delta_n$$

belongs to $S(E,R)$ and $a + \{x \in E : f(x) > 0\} \subset A$, which means A is $S(E,R)$-open. Hence, E is S-smooth.

An immediate corollary of this theorem is the following.

(5.2.10) 1°. <u>Every separable Banach space is</u> D_H-<u>smooth</u>.

2°. <u>Every separable Banach space with separable conjugate is</u> C^1-<u>smooth</u>.

3°. (c_0) <u>is</u> C^∞-<u>smooth</u>.

4°. L^p <u>and</u> ℓ^p (p>1) <u>are</u> C^∞-<u>smooth if</u> p <u>is an even integer and</u> $C^{[p]-1}$-<u>smooth otherwise</u>.

PROOF. 1° follows from (4.1.7), 2° from (4.2.5), 3° from (4.4.7) and 4° from (4.5.1).

The case of L^1 and ℓ^1 are covered by the following theorem.

(5.2.11) <u>If a Banach space</u> E <u>is separable and</u> $\tilde{E}$ <u>is not separable, then</u> E <u>is not</u> D-<u>smooth</u>.

For the proof of this important result of Whitfield [1], we refer the reader to Leach-Whitfield [1].

Some applications of (5.2.7) are the following.

(5.2.12) <u>Every separable</u> LCS <u>is</u> D_H-<u>smooth</u>.

PROOF. Every $E \in$ LCS is topologically isomorphic to a subspace of a topological product of Banach spaces $\{E_\sigma\}$. The separability of E implies that each E_σ is separable. Therefore, by (5.2.10). 1°, each E_σ is D_H-smooth.

Hence, by (5.2.7), E is D_H-smooth.

(5.2.13) <u>Every nuclear space is</u> EC^∞<u>-smooth</u>.

PROOF. Every nuclear space E is topologically isomorphic to a subspace of a topological product of $\{E_\sigma\}$, where E_σ is a subspace of ℓ^2. The space ℓ^2 is strongly EC^∞-smooth and, hence, each E_σ is EC^∞-smooth. Thus, by (5.2.7), E is EC^∞-smooth.

(5.2.14) <u>Every LCS with the weak topology is</u> EC^∞<u>-smooth</u>.

PROOF. The space in question is a subspace of a topological product of R (hence, nuclear).

(5.2.15) <u>Every</u> (FM)-<u>space is</u> EC^1<u>-smooth</u>.

PROOF. Since every (FM)-space is separable, by (4.1.7), it is strongly D_H-smooth. Since $B(E) = K_s(E)$, it is strongly D-smooth. Therefore, by (4.2.2) and (5.2.9), it is EC^1-smooth.

NOTE: In this section again, the main ideas and theorems are due to Bonic-Frampton [2], where a Banach space E was defined to be S-smooth if S(E,R) contains a non-trivial function with bounded support. Further, they have proved that E ∈ BS is S-smooth if and only if the topology is equivalent to the S(E,R)-topology. In the case of non-normed spaces, we can not follow the same route, because the existence of a continuous function with bounded support implies the existence of a bounded open set. Thus, we have let the S(E,R)-topology play the central rôle.

Theorem (5.2.11) is the only existing result which connects smoothness directly with the structure of the spaces. Other smoothness criterions are via strong smoothness. Theorems (5.2.7), (5.2.12) - (5.2.15) are due to Lloyd [2,3], where, as we have explained in the note of §5.1, the situation is slightly different.

§5.3 Partition of unity

In this section, we assume that S is a continuous S-category, $E \in LCS$ and $M \in \mathcal{O}(E)$.

The purpose of this section is to investigate the relations among the S-smoothness of E and the three properties defined in the following three definitions.

[5.3.1] A family of functions $\{\phi_\sigma \in S(M,R)\}$ is said to be an S-<u>partition of unity on</u> M if

1°. $\phi_\sigma(x) \geq 0$ for every σ and $x \in M$;

2°. the family $\{[\phi_\sigma]\}$ is a locally finite open covering of M, where $[\phi_\sigma] = \{x \in M : \phi_\sigma(x) > 0\}$;

3°. $\Sigma \phi_\sigma(x) = 1$ for each $x \in M$.

Let $\{U_\sigma\}$ be an open covering of M. An S-partition of unity $\{\phi_\sigma\}$ is said to be <u>subordinate</u> to $\{U_\sigma\}$ if

$$[\phi_\sigma] \subset U_\sigma \quad \text{for every } \sigma.$$

M is said to admit partitions of unity if, for any open covering $\{U_\sigma\}$ of M, there is a partition of unity which is subordinate to $\{U_\sigma\}$.

[5.3.2] M is said to be S-<u>normal</u> if, for any disjoint closed subsets H_0 and H_1 of M, there is $\phi \in S(M,R)$ such that $0 \leq \phi \leq 1$, $\phi(H_0) = 0$ and $\phi(H_1) = 1$.

[5.3.3] M is said to have <u>the</u> S-<u>approximation property</u> if, for any $F \in LCS$, $p \in P(F)$, $f \in C(M,F)$ and $\varepsilon > 0$, there exists $g \in S(M,F)$ such that

$$p(f(x) - g(x)) < \varepsilon \quad \text{for all } x \in M.$$

We have the following relations.

(5.3.4) <u>If</u> M <u>is normal and admits</u> S-<u>partitions of unity, then</u> M <u>is</u> S-<u>normal</u>. <u>If</u> M <u>is paracompact, the converse is true</u>.

PROOF. By the assumption, we can take $A \in \mathcal{O}(E)$ such that $H_0 \subset A \subset \bar{A} \subset M \setminus H_1$. Then, $\{A, M \setminus H_0\}$ is an open covering of M. Hence, there is an S-partition of unity $\{\phi_0, \phi_1\}$ which is subordinate to this covering; i.e., $[\phi_0] \subset A$ and $[\phi_1] \subset M \setminus H_0$. Then, ϕ_1 is the required function.

Conversely, assume that M is S-normal, paracompact and $\{U_\sigma\}$ be an open covering. By the paracompactness, we can assume that $\{U_\sigma\}$ are locally finite, and also $U_\sigma \subset M$. Since M is normal, we can take open sets $V_\sigma \subset M$ such that $\bar{V}_\sigma \subset U_\sigma$ for all σ and their union covers M. By the assumption, there are $\phi_\sigma \in S(M,R)$ such that $0 \leq \phi_\sigma \leq 1$, $\phi_\sigma(\bar{V}_\sigma) = 1$ and $\phi_\sigma(M \setminus V_\sigma) = 0$.

For each $x \in M$, there are only finitely many ϕ_σ such that $\phi_\sigma(x) \neq 0$. Therefore, we have $\phi : M \to R$ defined by

$$\phi(x) = \Sigma \phi_\sigma(x) \quad \text{for each } x \in M.$$

To prove that $\phi \in S(M,R)$, take an arbitrary $x \in M$. There exists a neighbourhood U of x such that $U \subset M$. There are finitely many σ_i such that $U \cap U_\sigma = \phi$ if $\sigma \neq \sigma_i$ for some i. Thus,

$$\phi/U = \Sigma \phi_{\sigma_i}/U \in S(M,R).$$

Hence, by [5.1.1].S.3, $\phi \in S(M,R)$.

Now, put

$$\psi_\sigma(x) = \phi_\sigma(x)/\phi(x) \quad \text{for each } \sigma \text{ and } x \in M.$$

Then, in a manner similar to above, we can show that $\psi_\sigma \in S(M,R)$, and $\{\psi_\sigma\}$ is an S-partition of unity on M subordinate to $\{U_\sigma\}$.

(5.3.5) If M admits S-partition of unity, then M has the S-approximation property.

PROOF. Let $f \in C(M,F)$, $p \in P(F)$ and $\varepsilon > 0$. For each $y \in F$, put

$$V_y = \{z \in F : p(y-z) < \varepsilon/2\},$$

and put $U_y = f^{-1}(V_y)$. Then, $\{U_y : y \in f(M)\}$ is an open covering of M. Let $\{\phi_y\}$ be an S-partition of unity which is subordinate to this covering. Since we may assume that $\phi_y \not\equiv 0$ for any y, we can take $x_y \in M$ such that $\phi_y(x_y) > 0$ for each $y \in f(M)$. Moreover, $\phi_y(x) > 0$ implies $p(y-f(x)) < \varepsilon/2$.

Now, put

$$g(x) = \sum_{y \in f(M)} f(x_y)\phi_y(x).$$

To show that $g \in S(M,F)$ by using [5.1.1]. S.3, take an arbitrary $x \in M$. Since $\{[\phi_y]\}$ is locally finite, $\phi_y(x) \neq 0$ for only finite numbers of $\{\phi_y\}$, which means that the above sum is locally a finite sum. Therefore, $g \in S(M,F)$. Moreover,

$$p(f(x)-g(x)) \leq \Sigma |\phi_y(z)| p(f(x_y) - f(x))$$

$$\leq \Sigma |\phi_y(x)| [p(f(x_y)-y) + p(y-f(x))] < \varepsilon.$$

(5.3.6) <u>If</u> M <u>is normal and has the</u> S-<u>approximation property, then</u> M <u>is</u> S-<u>normal</u>.

PROOF. Let H_0 and H_1 be disjoint closed sets. Since M is normal, there is $\psi \in C(M,R)$ such that $\psi(H_0) = 0$ and $\psi(H_1) = 1$. By the assumption, we can take $\phi \in S(M,R)$ such that

$$|\psi(x) - \phi(x)| < \frac{1}{4} \quad \text{for all} \quad x \in M.$$

Then, for $\mu \in C^\infty(R,R)$ such that

$$\mu(\xi) = 0 \text{ if } |\xi| \leq 1/3; \ = 1 \text{ if } 2/3 \leq \xi \leq 4/3,$$

the function $\phi_0 = \mu \circ \phi$ is the required function.

(5.3.7) <u>If</u> E <u>is</u> S-<u>normal, then</u> E <u>is</u> S-<u>smooth</u>.

PROOF. Obvious from (5.2.3).

(5.3.8) **If** E **is** S-smooth **and** M **has the** Lindelöf **property, then** M **admits** S-**partitions of unity**.

PROOF. Let $\{U_\sigma\}$ be an open covering of M. Since E is S-smooth, for any $x \in U_\sigma$, there exists $\phi \in S(E,R)$ such that $0 \leq \phi \leq 1$, $\phi(x) > 0$ and $[\phi] \subset U_\sigma$. In other words, each U_σ is the union of open sets of the form $[\phi]$ for such $\phi \in S(M,R)$.

Since M has the Lindelöf property, there is a refinement $\{W_i\}$ of $\{U_\sigma\}$ such that $W_i = [\phi_i]$, where $\phi_i \in S(M,R)$, $0 \leq \phi_i \leq 1$ and $\phi_i \neq 0$ for $i = 1,2,..$

Now, put $V_1 = W_1$ and

$$V_{n+1} = \{x \in M : \phi_{n+1}(x) > 0 \text{ and } \phi_i(x) < 1/n \text{ if } 1 \leq i \leq n\}.$$

Then, each V_n is open, and $\{V_n\}$ covers M.

In fact, take any $x \in M$. If $x \notin W_1$, then, take the smallest n such that $x \in W_n$. Then, $x \in V_n$. We denote such n by $n(x)$.

Let $\alpha : R \to R$ be such that $0 < \alpha(x) < \phi_{n(x)}(x)$ and put

$$V_x = \{z \in M : \phi_{n(z)}(z) > \alpha(x)\}.$$

Then, V_x is an open neighbourhood of x and $V_x \cap V_n = \phi$ if n is sufficiently large. Choose $\mu_{n+1} \in C^\infty(R^{n+1}, R)$ such that

$$\mu_{n+1}(\xi_1, \xi_2, \ldots, \xi_{n+1}) > 0 \text{ if and only if } \xi_i \leq 1/n \ (i=1,2,\ldots,n)$$

$$\text{and } \xi_{n+1} > 0.$$

Then,

$$\psi_{n+1} = \mu_{n+1} \circ (\phi_1 \times \ldots \times \phi_{n+1}) \circ \Delta_{n+1} \in S(M,R)$$

and

$$V_{n+1} = [\psi_{n+1}].$$

Now, put $\psi = \Sigma \psi_n$, which is a finite sum at each point and belongs to $S(M,R)$. Moreover, since $\psi_{n(x)}(x) > 0$, $\psi(x) > 0$ at every $x \in M$. Thus, since the inversion in R^+ is C^∞, $\psi^{-1}\psi_n \in S(M,R)$, where $(\psi^{-1}\psi_n)(x) = \psi_n(x)/\psi(x)$ for all $x \in M$. Then, $\{\psi^{-1}\psi_n\}$ is the required partition of unity.

NOTE: In this section, we have followed Lloyd [3].

We have assumed that S is continuous to ensure that, for any $\phi \in S(E,R)$, the set $[\phi]$ is open. It is possible to carry out most of the argument without this assumption. However, such an argument may not have any meaning from the topological point of view.

Goodman [1] has given equivalent results for the case which is roughly $S = D_H$ on Banach spaces. He first proved that separable Banach spaces have the D_H-approximation property, from which he has deduced the D_H-normality and the existence of D_H-partition of unity for separable Banach spaces.

There is no doubt that the most natural place to develop the theory of this section is on a manifold modelled on LCS. In fact, Lloyd [3] has presented such a theory.

Bolis [1] also treated the same problem, but by a different way. Instead of employing a particular kind of differentiation, he defined a differentiation by the properties which are satisfied by such differentiations as those of Bastiani (IV in §1.12) and Frölicher-Bucher [1]. Using the notion C^n defined in this way, he first proved the C^n-normality of strongly C^n-smooth separable LCS, and then proved the existence of C^n-partition of unity for paracompact C^n-manifolds modelled on a strongly C^n-smooth separable MLS.

Another interesting work by Leduc [1] has also treated the same problem for Banach spaces.

CHAPTER 6. DIFFERENTIABILITY OF MAPPINGS

OF A REAL VARIABLE

When one has to show that a given mapping is differentiable, the first thing to do is to investigate its directional differentiability, and, if it is directionally differentiable, then the theorems in §1.4 can be used to investigate the stronger differentiability.

Let $E, F \in$ LCS, $A \in \mathcal{O}(E)$ and $f : A \to F$. Then, the directional differentiability of f at $a \in A$ in the direction of $x \in E$ is equivalent to the differentiability of the mapping $f(a+\xi x)$ with respect to ξ at $\xi = 0$. In other words, the basic case is that of $f : A \to E$, where $A \in \mathcal{O}(R)$ and $E \in$ LCS.

We can simplify the problem further by taking $\bar{x} \in \bar{E}$ and considering the function $\phi : A \to R$ defined by

$$\phi(\xi) = < f(\xi), \bar{x} > .$$

There are several criteria for the differentiability of a real-valued function of a real variable, among which the most fundamental is the theorem of Denjoy-G.C. Young-Saks (see Saks [1,p.271]) or, equivalently, that of Stepanoff (see §6.2) The conclusions of these theorems are almost everywhere (a.e.) differentiability.

In other words, the best result we can reasonably expect for ϕ defined above is differentiability except for a null set which depends upon the choice of $\bar{x}$. Since the number of $\bar{x}$ is more than countable in general, the problem of stepping up from here to the a.e. differentiability of f itself is far from simple.

§6.1 Differentiability of Lipschitz mappings

Let $I = [\alpha, \beta]$ and $E \in$ BS. A mapping $f : I \to E$ is called a Lipschitz mapping if there exists $\gamma > 0$ such that

$$\|f(\xi) - f(\eta)\| \leq \gamma |\xi - \eta| \quad \text{if} \quad \xi, \eta \in I.$$

It is a well-known theorem of Rademacher that, if E is finite-dimensional, f is differentiable a.e.. In general, this is not true.

Example. Define a function $f : [0,1] \to L^1[0,1]$ by

$$f(\xi)(\eta) = 1 \text{ if } 0 \leq \eta \leq \xi; = 0 \text{ otherwise.}$$

Then, if $\xi_1 < \xi_2$ in $[0,1]$,

$$\|f(\xi_1) - f(\xi_2)\| = \int_0^1 |f(\xi_1)(\eta) - f(\xi_2)(\eta)| d\eta = \int_{\xi_1}^{\xi_2} d\eta = \xi_2 - \xi_1.$$

Therefore, f is a Lipschitz mapping on $[0,1]$. Assume that $f'(\xi_0)$ exists for $\xi_0 \in (0,1)$. Then, for any $g \in M[0,1]$, the following limit exists :

$$\lim_{\varepsilon \to 0} \varepsilon^{-1} \int_0^1 [f(\xi_0+\varepsilon) - f(\xi_0)](\eta) g(\eta) d\eta = \lim_{\varepsilon \to 0} \varepsilon^{-1} \int_{\xi_0}^{\xi_0+\varepsilon} g(\eta) d\eta.$$

However, this limit does not exist if, for instance, we take g which is 1 if $\eta > \xi_0$ and -1 if $\eta \leq \xi_0$. This example is due to Gelfand [1].

For the further study, the following three notions are fundamental.

[6.1.1] $f : I \to E$ is said to be <u>separably valued</u> if the closed linear subspace spanned by $f(I)$ is separable.

[6.1.2] A sequence $\{\bar{x}_n\}$ is said to be (*)-<u>weakly dense</u> if $\|\bar{x}_n\| \leq 1$ and, for any $\bar{x} \in \bar{E}$ such that $\|\bar{x}\| \leq 1$, there is a subsequence $\{\bar{x}_{n_i}\}$ such that

$$\lim_{i \to \infty} < x, \bar{x}_{n_i} > = < x, \bar{x} > \text{ for any } x \in E.$$

It is known that, if E is separable, there is always a (*)-weakly dense sequence in $\bar{E}$ (Yoshida [1], p.131-132). Moreover, if $\{\bar{x}_n\}$ is (*)-weakly dense, it is easily seen that

$$\|x\| = \overline{\lim_{n \to \infty}} |< x, \bar{x}_n >| \text{ for any } x \in E.$$

[6.1.3] $f : I \to E$ is said to be $\{\bar{x}_n\}$-<u>pseudodifferentiable</u> to g if $\{\bar{x}_n\}$ is (*)-weakly dense and, for each n,

$$\frac{d}{d\xi} < f(\xi), \bar{x}_n > = < g(\xi), \bar{x}_n >$$

for all $\xi \in I$ except for a null set (which may depend on $\bar{x}_n$).

In the following we shall use some elementary properties of the Bochner integral. For these, see Yoshida [1], p.130-136. The first fundamental tool is the following theorem of Pettis [1]. The proof is the same as the second half of the proof on p.131 of Yoshida [1], where it was proved that <u>a mapping is measurable if and only if it is weakly measurable and a.e. separably valued</u>.

(6.1.4) <u>Let</u> E <u>be separable and</u> $\{\bar{x}_n\}$ <u>be</u> (*)-<u>weakly dense. Let</u> $f : I \to E$. <u>If the functions</u> $< f(\xi), \bar{x}_n >$ <u>are measurable for all</u> n, <u>then</u> f <u>is measurable</u>.

Next, we shall prove the differentiability of Lipschitz mappings which will be used in the next section.

(6.1.5) <u>Let</u> $f : I \to E \in BS$ <u>be a Lipschitz mapping. If</u> f <u>is</u> $\{\bar{x}_n\}$-<u>pseudo-differentiable to a separably valued</u> g, <u>then</u> f <u>is a.e. differentiable to</u> g.

PROOF. Since f is continuous, $f(I)$ is separable. Therefore, since g is also separably valued, we may suppose that E itself is separable. We shall first prove that g is integrable. To this end, it is sufficient to show that g is measurable and $\|g(\xi)\|$ is integrable. The first follows from (6.1.4), because each $< g(\xi), \bar{x}_n >$, being the limit of continuous functions $< f_\varepsilon(\xi), \bar{x}_n >$, where $f_\varepsilon(\xi) = \varepsilon^{-1}[f(\xi+\varepsilon) - f(\xi)]$ for $\varepsilon \neq 0$, is measurable. The second follows from the fact that the measurable function $\|g(\xi)\|$ is bounded a.e., which can be proved as follows. Let N_n be the null sets for which

$$\frac{d}{d\xi} < f(\xi), \bar{x}_n > = < g(\xi), \bar{x}_n > \quad \text{if} \quad \xi \in I \setminus N_n.$$

Then, $N = \bigcup_{n=1}^{\infty} N_n$ is a null set and, if $\xi \in I \setminus N$,

$$\|g(\xi)\| = \overline{\lim_{n \to \infty}} |< g(\xi), \bar{x}_n >| = \overline{\lim_{n \to \infty}} \lim_{\varepsilon \to 0} |< f_\varepsilon(\xi), \bar{x}_n >| \leq \gamma,$$

where γ is the Lipschitz constant of f. Thus, we can define a function h : I → E by

$$h(\xi) = \int_\alpha^\xi g(\eta)d\eta + g(\alpha),$$

where we may assume that $g(\alpha) = f(\alpha)$. Then,

$$<h(\xi), \bar{x}_n> = \int_\alpha^\xi <g(\eta), \bar{x}_n> d\eta + <g(\alpha), \bar{x}_n>$$

$$= \int_\alpha^\xi \frac{d}{d\eta} <f(\eta), \bar{x}_n> d\eta + <f(\alpha), \bar{x}_n> = <f(\xi), \bar{x}_n>,$$

which implies $h(\xi) = f(\xi)$, because $\{\bar{x}_n\}$ is total. Thus, f is a.e. differentiable.

When E is reflexive, we have a simpler criterion.

(6.1.6) Let E be reflexive and f : I → E be a Lipschitz mapping. Then, f is differentiable a.e.

PROOF. We can again assume that E is separable. If $\xi \in (\alpha, \beta)$ is fixed, the set $\{f_\varepsilon(\xi) : \varepsilon \neq 0\}$ is bounded. Therefore, since E is reflexive, it is weakly relatively compact. Hence, we have g : $(\alpha, \beta) \to E$ such that

$$g(\xi) = \text{weak - } \lim_{\varepsilon \to 0} f_\varepsilon(\xi).$$

Thus, the assumption of (6.1.5) are satisfied.

NOTE: We are, in this and the next sections, following the route which has been started by Pettis [1] and improved by Alexiewicz [1]. See also the appendix of Komura [1].

§6.2 Differentiability of Stepanoff mappings.

We first define the Stepanoff mapping. Let A ⊂ R be a measurable set.

[6.2.1] f : A → E ∈ BS is called a Stepanoff mapping if

$$\overline{\lim_{\varepsilon \to 0}} \ \|f_\varepsilon(\xi)\| < \infty \qquad \text{for almost all } \xi \in A.$$

When E is finite-dimensional, as Stepanoff has shown, any Stepanoff mapping is a.e. differentiable. Note that this condition is equivalent to all Dini derivatives being finite a.e. In general, Stepanoff mappings are not always differentiable a.e. as the example before [6.1.1] shows.

Before proceeding to the differentiability problem, we shall clarify the position of these mappings among other important classes of mappings.

A mapping $f : I \to E$ is said to be <u>of bounded variation</u> if

$$\text{var}(f) = \sup_\pi \sum_{i=0}^{n} \|f(\xi_i) - f(\xi_{i-1})\| < \infty$$

where the supremum is taken over all partitions $\pi = (\alpha = \xi_0, \xi_1, \ldots, \xi_n = \beta)$ of I.

A mapping $f : I \to E$ is said to be <u>absolutely continuous</u> if for any $\varepsilon > 0$ there exists $\delta > 0$ such that the following condition is satisfied: for any partition $\pi = (\alpha = \xi_0, \xi_1, \ldots, \xi_n = \beta)$ such that $\sum_{i=1}^{n} |\xi_i - \xi_{i-1}| < \delta$ implies $\sum_{i=1}^{n} \|f(\xi_i) - f(\xi_{i-1})\| < \varepsilon$.

It is obvious that Lipschitz mappings are absolutely continuous, which implies they are of bounded variation.

(6.2.2) <u>If</u> $f : I \to E \in BS$ <u>is of bounded variation, then</u> f <u>is a Stepanoff mapping.</u>

PROOF. Put $\phi(\xi) = \lim_{\varepsilon \to 0+} \|f_\varepsilon(\xi)\|$. Since ϕ is measurable, the set $\{\xi : \phi(\xi) = \infty\}$ is measurable. Assume that

$$\lambda = |\{\xi : \phi(\xi) = \infty\}| > 0.$$

Put $A_n = \{\xi \in I : \sup_{\varepsilon \geq 1/n} \|f_\varepsilon(\xi)\| \geq \frac{2}{\lambda} \text{var}(f)\}$. Then, each A_n is measurable and

$$\bigcup_{n=1}^{\infty} A_n \supset \{\xi : \phi(\xi) = \infty\}.$$

Since $\{A_n\}$ is increasing, there is n such that $|A_n| \geq 2/\lambda$.

Define $\{\xi_i\}$ and $\{\varepsilon_i\}$ as follows :

$$\xi_1 = \inf_{\xi \in A_n} \xi, \quad \xi_{i+1} = \inf \{\xi \in A_n : \xi \geq \xi_i + \varepsilon_i\}$$

and

$$\varepsilon_i = \sup \{\varepsilon : \|f_\varepsilon(\xi_i)\| \geq \frac{2}{\lambda} \text{var}(f)\}.$$

Then,

$$\bigcup_{i=1}^{\infty} [\xi_i, \xi_i + \varepsilon_i] \supset A_n.$$

Hence,

$$\Sigma \|f(\xi_i + \varepsilon_i) - f(\xi_i)\| \geq \frac{2}{\lambda} \text{var}(f) \Sigma \varepsilon_i \geq \frac{2}{\lambda} \text{var}(f) |A_n| > \text{var}(f),$$

which is a contradiction. Similarly, $\varlimsup_{\varepsilon \to 0-} \|f_\varepsilon(\xi)\|$ is finite a.e.. Thus, f is a Stepanoff mapping.

Thus, we have the following implications :

Lipschitz → A.C. → B.V. → Stepanoff.

The following theorem shows that a Stepanoff mapping is, in a sense, a collection of Lipschitz mappings.

(6.2.3) Let $f : A \to E \in BS$ be a Stepanoff mapping. Then, A can be a.e. decomposed into a sequence of measurable subsets on each of which f is Lipschitzian.

PROOF. For $(\xi, \delta) \in A \times R^+$, define

$$Q(\xi, \delta, n) = (\xi - \delta, \xi + \delta) \cap \{\eta \in A : \|f(\xi) - f(\eta)\| > n|\xi - \eta|\}.$$

Since f is continuous a.e., $Q(\xi, \delta, n)$ is a measurable subset of $A \times R^+$. Put

$$A_n = \{\xi \in A : |Q(\xi, \delta, n)| < \delta/2 \text{ if } 0 < \delta < 1/n\}.$$

By Fubini's theorem, $|Q(\xi,\delta,n)|$ is a measurable function of ξ. Hence, A_n is measurable and $A = \bigcup_{n=1}^{\infty} A_n$.

Further, if $\xi, \eta \in A_n$ and $|\xi-\eta| < 1/n$, we have

$$\|f(\xi) - f(\eta)\| \leq 2n|\xi-\eta|.$$

In fact, from the definition of A_n,

$$|Q(\xi,|\xi-\eta|,n) \cup Q(\eta,|\xi-\eta|,n)|$$

$$\leq |\xi-\eta| = |B(\xi,|\xi-\eta|) \cap B(\eta,|\xi-\eta|)|.$$

By the definition of the set Q, we can choose ρ such that

$$\rho \in B(\xi,|\xi-\eta|) \cap B(\eta,|\xi-\eta|) \setminus Q(\xi,|\xi-\eta|,n) \cup Q(\eta,|\xi-\eta|,n).$$

Then,

$$\|f(\xi) - f(\eta)\| \leq \|f(\xi) - f(\rho)\| + \|f(\rho) - f(\eta)\| \leq n|\xi-\rho| + n|\rho-\eta|$$

$$\leq n|\xi-\eta| + n|\xi-\eta| = 2n|\xi-\eta|.$$

Thus, we only have to express each A_n as the union of measurable subsets with diameters less than $1/n$.

Now, we prove the main theorem of this section.

(6.2.4) Let $f : I \to E \in BS$ be a Stepanoff mapping which is $\{\bar{x}_n\}$-pseudo-differentiable to a separably valued g. Then, f is a.e. differentiable to g.

PROOF. By (6.2.3), I is an a.e. union of countable measurable subsets on each of which f is Lipschitzian. Assume that f is Lipschitzian on a measurable subset $A \subset I$. Since A can be approximated by closed sets from the inside, we may assume that A itself is closed. Let $\{(\alpha_n,\beta_n)\}$ be a sequence of open intervals in (α,β) which are contiguous to A. Define $h : (\alpha,\beta) \to E$ by

$$h(\xi) = f(\xi) \text{ if } \xi \in A; = f(\alpha_n) + (\beta_n-\alpha_n)^{-1}(f(\beta_n)-f(\alpha_n))(\xi-\alpha_n) \text{ if } \xi \in (\alpha_n,\beta_n).$$

Then, h is Lipschitzian and $\{\bar{x}_n\}$-pseudodifferentiable to a separably valued mapping. Hence, by (6.1.5), h is a.e. differentiable. Therefore, f is a.e. differentiable on A, which implies f is differentiable a.e. on I.

In the proof above, the $\{\bar{x}_n\}$-pseudodifferentiability was used only to imply the a.e. differentiability of Lipschitz mappings. Therefore, by (6.1.6), we have the following theorem.

(6.2.5) Let E be reflexive. Then, any Stepanoff mapping of I into E is a.e. differentiable.

Another interesting application of (6.2.4) is the following.

(6.2.6) If $f : I \to E \in BS$ is weakly differentiable to g, then f is a.e. differentiable to g.

PROOF. Since the assumption means that the limit

$$\lim_{\varepsilon \to 0} < f_\varepsilon(\xi), \bar{x} >$$

exists for each $\xi \in I$ and each $\bar{x} \in \bar{E}$, f is a Stepanoff mapping. Thus, the conditions of (6.2.4) are satisfied.

Here we add two convenient criteria for differentiability in the case when $E = R$.

The first, which is due to Khintichine [1], implies that symmetric differentiability implies a.e. differentiability. In Saks [1], p.151, this was proved in a more general situation. The proof below is Khinchine's original one.

(6.2.7) Let $f : I \to R$ be measurable and

(#) $\quad \overline{\lim_{\varepsilon \to 0}} (2\varepsilon)^{-1}[f(\xi+\varepsilon) - f(\xi-\varepsilon)] < \infty \quad$ for $\xi \in I$ a.e.

Then, f is a.e. differentiable.

PROOF. Let A be the set of points where (#) is true and f is not differentiable. Assume that $|A| > 0$. Put

$$A_n = \{\xi \in A : f(\xi+\epsilon) - f(\xi-\epsilon) < 2\epsilon n \text{ if } 0 < \epsilon < 1/n\}.$$

Then, $A = \bigcup_{n=1}^{\infty} A_n$. Then, $|A_{n_0}| > 0$ for some n_0. Put

$$\phi(\xi) = f(\xi) - n_0\xi.$$

Then, if $\xi \in A_{n_0}$ and $0 < \epsilon < 1/n_0$, we have

(*) $\qquad \phi(\xi+\epsilon) - \phi(\xi-\epsilon) < 0.$

We shall prove that the Dini upper derivatives $\phi^+(\xi)$ and $\phi^-(\xi)$ are negative if ξ is a point of outer density of A_{n_0}.

To do this, let ξ be a point of outer density of A_{n_0}. Then, there is $\delta > 0$ such that $\delta < 1/n_0$ and

(⁎) $\qquad |E_{n_0} \cap (\xi,\eta)| > \frac{3}{4}(\eta-\xi) \quad \text{if } \xi < \eta < \xi + \delta.$

Suppose that $\phi(\xi) \leq \phi(\eta)$ for such η. Put

$$M_1 = \{\rho \in (\xi,\eta) : \phi(\rho) \geq \phi(\xi)\}, \quad M_2 = \{\rho \in (\xi,\eta) : \phi(\rho) < \phi(\xi)\}$$

and

$$M_3 = \{\tfrac{1}{2}(\rho+\xi) : \rho \in M_1\}, \quad M_4 = \{\tfrac{1}{2}(\rho+\eta) : \rho \in M_2\}.$$

These sets are measurable and the measure of one of M_1 and M_2 is at least $\frac{1}{2}(\eta-\xi)$. Hence, the measure of one of M_3 and M_4 is at least $\frac{1}{4}(\eta-\xi)$. Therefore, by (⁎), one of M_3 and M_4 has non-empty intersection with A_{n_0}.

Suppose that $\rho_0 \in M_3 \cap A_{n_0}$. Then, since $\rho_0 + \rho_0 - \xi \in M_1$,

$$\phi(\rho_0+(\rho_0-\xi)) \geq \phi(\xi) = \phi(\rho_0-(\rho_0-\xi)),$$

and

$$0 \leq \rho_0 - \xi \leq \eta - \xi \leq \delta \leq 1/n_0,$$

which contradicts (*).

Suppose that $\rho_0 \in M_4 \cap A_{n_0}$. Then, since $\rho_0 + \rho_0 - \eta \in M_2$,

$$\phi(\rho_0-(\eta-\rho_0)) < \phi(\xi) \leq \phi(\eta) = \phi(\rho_0+(\eta-\rho_0))$$

and

$$0 \leq \eta - \rho_0 \leq \eta - \xi \leq \delta \leq 1/n_0,$$

which again contradicts (*).

Thus, we have $\phi(\xi) > \phi(\eta)$ if $\xi < \eta < \xi + \delta$, which implies that $\phi^+(\xi)$ is negative. Similarly, $\phi^{-1}(\xi)$ is negative. Since these relations are true at any point of outer density of A_{n_0}, they are true a.e. in A_{n_0}. Hence, by Denjoy-G.C.Young-Saks theorem (Saks [1], p.271), f is a.e. differentiable in A_{n_0}, which is a contradiction. Therefore, f is a.e. differentiable in I.

Another criterion, due to Zygmund [1], is in a sense the dual to the above theorem.

(6.2.8)   Let f : I → R be continuous and

(#) $\lim_{\varepsilon \to 0} \varepsilon^{-1}[f(\xi+\varepsilon) + f(\xi-\varepsilon) - 2f(\xi)] = 0$ for every $\xi \in (\alpha,\beta)$.

Then, f is differentiable in an everywhere dense subset of I.

PROOF. Firstly, if $f(\xi)$ has maximum or minimum at $\xi \in (\alpha,\beta)$, then $f'(\xi)$ exists and equals to zero. In fact, since

$$\varepsilon^{-1}[f(\xi+\varepsilon) + f(\xi-\varepsilon) - 2f(\xi)] = \varepsilon^{-1}[f(\xi+\varepsilon) - f(\xi)] + \varepsilon^{-1}[f(\xi-\varepsilon) - f(\xi)],$$

for $\varepsilon > 0$, $\varepsilon^{-1}[f(\xi+\varepsilon) - f(\xi)]$ and $\varepsilon^{-1}[f(\xi-\varepsilon) - f(\xi)]$ have the same sign. Therefore, by (#), they have to converge to zero.

Now, let $(\alpha_1,\beta_1) \subset (\alpha,\beta)$ and let

$$g(\xi) = \lambda\xi + \mu$$

be a linear function such that

$$g(\alpha_1) = f(\alpha_1) \quad \text{and} \quad g(\beta_1) = f(\beta_1).$$

Then, again, the function $h = f - g$ satisfies (#) and $h(\alpha_1) = h(\beta_1) = 0$. Since h is continuous, it has a maximum or minimum at some $\xi \in (\alpha_1, \beta_1)$. Hence, $h'(\xi) = 0$ and $f'(\xi) = \lambda$.

NOTE: As to (6.2.3) and (6.2.4), we have followed Federer [1]. The theorems (6.2.5) and (6.2.6) have been obtained by Pettis [1] and Alxiewicz [1]. Daneš-Kolomý [1] has an interesting use of the condition (#) in (6.2.8)

§6.3 Results of L. Schwartz and Grothendieck

The results in this section are concerned with the relation between continuous weak differentiability and continuous differentiability.

Throughout, let $f : R \to E \in LCS$.

[6.3.1] Let T_1 and T_2 be two LC-topologies on E. Then, T_1 and T_2 are said to be (P)-<u>related</u> if T_2 is weaker than T_1 and, for any $U \in N(E, T_1)$, there exists $V \in N(E, T_1)$ such that the convex T_2-closure of any T_1-compact subset of V is contained in U.

Let T_1 be the original topology of E and T_2 be the weak topology. Then, since closed convex sets are weakly closed, these are (P)-related.

(6.3.2) <u>Let T_1 and T_2 be (P)-related. If</u> $f \in C^n(R, E)$ <u>by</u> T_2 <u>and</u> $f^{(n)}$ <u>is continuous by</u> T_1, <u>then</u> $f \in C^n(R, E)$ <u>by</u> T_1.

PROOF. We shall only prove the case when $n = 1$. Let $\alpha \in R$. Since f' is continuous at α by T_1, for any $U \in N(E, T_1)$, there is $\varepsilon_0 > 0$ such that $|\varepsilon| < \varepsilon_0$ implies $f'(\alpha+\varepsilon) - f'(\alpha) \in V$, where $V \in N(E, T_1)$ is the one determined from the definition [6.3.1]. On the other hand, since f is differentiable by T_2, by the mean value theorem,

$$\varepsilon^{-1}[f(\alpha+\varepsilon) - f(\alpha)] - f'(\alpha) \in |\overline{Co}|\{f'(\xi) - f'(\alpha) : \xi \in (\alpha, \alpha+\varepsilon)\},$$

and the right-hand side is, when $0 < |\varepsilon| < \varepsilon_0$, a convex T_2-closure of a T_1-compact subset of V. Therefore,

$$\varepsilon^{-1}[f(\alpha+\varepsilon) - f(\alpha)] - f'(\alpha) \in U \quad \text{if} \quad 0 < |\varepsilon| < \varepsilon_0,$$

which means that f is differentiable at α by T_1.

In particular, if $f \in C^n(R,E)$ weakly and $f^{(n)}$ is strongly continuous, then $f \in C^n(R,E)$ by the original topology.

In the following theorem, strong continuity is replaced by weak continuity but with a weaker conclusion.

(6.3.3) <u>Let</u> E <u>be quasi-complete and</u> $f \in C^n(R,E)$ <u>in the weak topology</u>. <u>Then</u>, $f \in C^{n-1}(R,E)$ <u>in the original topology</u>.

PROOF. Assume that $n = 1$ and, for $\alpha \in R$, put

$$B = \{|\varepsilon|^{-1}[f(\alpha+\varepsilon) - f(\alpha)] : 0 < |\varepsilon| \leq 1\}.$$

By assumption this is weakly bounded and hence bounded. Thus f is continuous at α.

Let $n = 2$. By Köthe [1], p.297-298, the dual of $\bar{E}$ is a subspace of its algebraic dual $\bar{E}^*$, which can be regarded as the weak completion of E, and $\bar{E}$ is the union of the weak closures in $\bar{E}^*$ of bounded subsets of E.

By the assumption, $f' : R \to \bar{E}^*$. On the other hand, $f'(\alpha)$ is the weak limit of $\{f_\varepsilon(\alpha)\}$, which, if $0 < |\varepsilon| < \varepsilon_0$, is contained in a bounded subset of E. Therefore, $f'(\alpha) \in \bar{\bar{E}}$.

Now, since the set $A = \{f(\alpha) : 0 < |\varepsilon| \leq 1\}$ is a set of continuous linear functionals on $\bar{E}$ and, since $f \in C^2(R,E)$ weakly, by the mean value theorem A is bounded in the topology T_2 of simple convergence on $\bar{E}$.

Let T_1 be the natural topology (Köthe [1], p.300), which is the topology of uniform convergence on the equicontinuous subsets of $\bar{E}$. Then, A is bounded in

T_1. Therefore, $f'(\xi)$ is T_1-continuous with respect to ξ.
Thus, by (6.3.2), f' is the T_1-derivative of f, i.e.,

$$f_\varepsilon(\alpha) \to f'(\alpha) \in \bar{\bar{E}} \quad \text{with} \quad T_1.$$

On the other hand, the natural topology induces the original topology on E. Therefore, the Cauchy filter A is bounded in E and, since E is quasi-complete, $f'(\alpha) \in E$ and it is the derivative of f by the original topology of E. That f' is continuous follows from the step 1 and $f \in C^2(R,E)$ weakly.

It can be seen from the above proof that it is enough to assume that $f \in C^{n-1}(R,E)$ weakly and $f^{(n-1)}$ is Lipschitzian.

NOTE: These theorems are due to L.Schwartz [1] and Grothendieck [1] respectively.

CHAPTER 7. SETS OF DIFFERENTIABLE MAPPINGS

Let S be an S-category and $E, F \in LCS$. Then, by (5.1.7), $S(E,F)$ is a linear space. Further, by the definition, $S(E) = S(E,E)$ is a semigroup by the composition. Since

$$(f+g) \circ h = f \circ h + g \circ h,$$

$S(E)$ is also a near-ring.

The properties of these algebraic structures have not been fully investigated and there are many unsolved problems. The aim of this chapter is to introduce some results which are related to the algebraic structures of the sets of differentiable mappings.

§7.1 Idempotents of the semigroups of differentiable mappings

Let $E \in LCS$ and S be an S-category. $f \in S(E)$ is said to be an idempotent if $f \circ f = f$, which is equivalent to say that f is equal to the identity on the image of f. The study of the idempotents in semigroups of differentiable mappings has been started by Nadler [1]. All results in this section are essentially due to Nadler [1,2].

(7.1.1) Let $I = [0,1]$ and $D(I)$ be the semigroup of all differentiable functions of I into itself. At the endpoints, the differentiability is to be the one-side differentiability. Then, $f \in D(I)$ is an idempotent if and only if it is either the identity or a constant function.

PROOF. Since the sufficiency is obvious, let us assume that $f \in D(I)$ is an idempotent and it is not constant. Then, $f(I) = [\alpha, \beta]$ for some $\alpha, \beta \in I$ such that $\alpha < \beta$.

If $I = [\alpha, \beta]$, since f is equal to the identity on its image, f itself is the identity.

Suppose that $I \neq [\alpha, \beta]$. If $\beta \neq 1$, then

$$f'(\beta) = \lim_{\varepsilon \to 0} \varepsilon^{-1} f_\varepsilon(\beta) = \lim_{\varepsilon \to 0} \varepsilon^{-1}[(\beta+\varepsilon) - \beta] = 1.$$

Therefore, f is increasing in a neighbourhood of β. Hence, if $\varepsilon > 0$ is small,

$$\beta = f(\beta) < f(\beta+\varepsilon) \quad \text{and} \quad \beta + \varepsilon \in I,$$

which contradicts the assumption that $f(I) = [\alpha,\beta]$.

Finally, assume that $\beta = 1$. Then, $\alpha \neq 0$ and $f'(\alpha) = 1$, which again leads to the same contradiction.

The following theorem, which is a special case of the main theorem of Nadler [2], is, in a sense, an infinite-dimensional version of (7.1.), although here we have to assume that the mappings involved are C^1. It is not known whether C^1 can be replaced by D. We denote the interior of a set M by int(M).

(7.1.2) Let $E \in BS$ and let $M \subset E$ be a connected open set such that $\text{int}(\bar{M}) = M$. Let $C^1(\bar{M})$ be the semigroup of all mappings of $\bar{M}$ into $\bar{M}$ each of which is C^1 in an open set containing $\bar{M}$. Then, for any idempotent f of $C^1(\bar{M})$ different from the identity, $f(\bar{M})$ has empty interior.

PROOF. Put $A = \text{int}(f(\bar{M}))$ and assume that A is not empty. We first show that

(*) $\bar{A} \cap M = A.$

Since $A \subset \bar{A} \cap M$ is obvious, assume that $x \in \bar{A} \cap M$. Then, $x_n \to x$ for some $x_n \in A$. Since f is equal to the identity on $f(\bar{M})$, $f(x_n) = x_n$, and, hence, $f(x) = x$. Moreover, since $f'(x_n) = 1$ and f is C^1, we have $f'(x) = 1$. By the inverse function theorem (3.4.6), there exist open neighbourhoods U and V of x such that $f : U \to V$ is a homeomorphism. Since $x \in M$, we may assume that $U \subset M$. Hence, $V = f(U) \subset f(M)$, which implies that $x \in \text{int}(f(\bar{M})) = A$.

Thus, we obtain (*), which show that $M = A \cup (M \setminus \bar{A})$. Moreover, it again follows from (*) that A and $M \setminus \bar{A}$ are disjoint. Since M is connected and both sets are open, we must have $M \setminus \bar{A}$ is empty, because we have assumed that A is not empty. Then, it follows that $M \subset \bar{A}$. Since f is equal to the identity on $\bar{A}$, it is equal to the identity on M and, hence, on $\bar{M}$, which is a contradiction.

When E is finite-dimensional, we have more detailed information.

(7.1.3) Let $C^1(M)$ be the same semigroup as above and let E be n-dimensional. Then, for any idempotent f of $C^1(\bar{M})$ which is different from the identity, we have

1°. dimension of $f(\bar{M}) \leq n - 1$;

2°. f is not locally one to one at any point of M.

PROOF. If the dimension of $f(\bar{M})$ is n, there is a non-empty subset of $f(\bar{M})$ which is open in E (Hurwitz-Wallman [1], p.44, Theorem IV.3). This contradicts the conclusion of (7.1.2).

2°. Let $x \in M$ and U be an open bounded set such that $x \in U$ and $\bar{U} \subset M$. Since the dimension of $\bar{U}$ and $f(\bar{U})$ are n and less than n - 1 respectively, $f/\bar{U}$ is a mapping of a compact set $\bar{U}$ into $\bar{M}$ which lowers dimension. By Hurwitz-Wallman [1], p.91, theorem VI.7, there exists $a \in \bar{M}$ such that the dimension of $(f/\bar{U})^{-1}(a)$ is not less than 1. Therefore, f is not one-to-one on any neighbourhood of x.

§7.2 Automorphisms of semigroups of differentiable mappings

An automorphism ϕ of a semigroup A is a bijection $\phi : A \to A$ that is multiplicative :

$$\phi(fg) = \phi(f)\phi(g) \quad \text{for all} \quad f,g \in A.$$

It is said to inner if there is an invertible $h \in A$ such that

(*) $\quad \phi(f) = hfh^{-1} \quad \text{for all} \quad f \in A.$

The aim of this section is to consider when an automorphism is inner.

Firstly, let E be any set and A be a semigroup of mappings of E into E, where the product fg for $f,g \in A$ is defined by the composition: $(fg)(x) = f(g(x))$ for $x \in E$. We denote the __constant mapping__ whose value is a by c_a. Hence,

$$c_a(x) = a \text{ for all } x \in E; \quad fc_a = c_{f(a)} \text{ and } c_a f = c_a.$$

We denote the set of all constant mappings by $I(E)$. The essential part of the following fact is due to Magill [1].

(7.2.1) __Let A be a semigroup of mappings of a set E into itself. If $I(E) \subset A$, then, for any automorphism ϕ of A, there exist a bijection h of E such that (*) holds.__

PROOF. We first prove that $\phi : I(E) \to I(E)$. Let $c_x \in I(E)$. For any $y \in E$ there exists $f \in A$ such that $\phi(f) = c_y$. Hence, for any $z \in E$,

$$\phi(c_x)(y) = \phi(c_x)\phi(f)(z) = \phi(c_x f)(z) = \phi(c_x)(z),$$

which shows that $\phi(c_x) \in I(E)$.

Thus, we can define $h : E \to E$ by

$$\phi(c_x) = c_{h(x)} \quad \text{for all } x \in E.$$

Then, h is one-to-one, because $h(x_1) = h(x_2)$ implies $\phi(c_{x_1}) = \phi(c_{x_2})$, which leads to $x_1 = x_2$.

Further, h is onto, because, for any $y \in E$, if we take $f \in A$ such that $\phi(f) = c_y$, we can show in the same way as above that $f \in I(E)$, or, $f = c_x$ for some $x \in E$, and, hence, $y = h(x)$.

Finally, let $f \in A$ and $x \in E$. Then, for any $x \in E$,

$$\phi(f)(x) = \phi(f)c_x(z) = \phi(f)(c_{h^{-1}(x)})(z) = \phi(fc_{h^{-1}(x)})(z)$$

$$= \phi(c_{fh^{-1}(x)})(z) = c_{hfh^{-1}(x)}(z) = hfh^{-1}(x),$$

which implies (*).

Now, let S be an S-category and E ∈ LCS. The semigroup $S(E) = S(E,E)$ contains $I(E)$, because every constant mapping is EC^∞. Therefore, any automorphism of $S(E)$ can be written in the form (*). Since the mapping h corresponds uniquely to ϕ, if we start with ϕ^{-1} instead of ϕ, we arrive at (*) with h^{-1} in the place of h. This shows that any statement that has been proved for h can be assumed to be true for h^{-1}.

We shall first prove the weak continuity of h. The essential part of the following proof is due to Magill [1].

(7.2.2) <u>Let</u> S <u>be a continuous</u> S-<u>category</u>, ϕ <u>be an automorphism of the semigroup</u> $S(E)$ <u>and</u> h <u>be the mapping in</u> (*). <u>Then, for any</u> $\bar{a} \in \bar{E}$, <u>the function</u> $< h(x), \bar{a} >$ <u>is continuous with respect to</u> $x \in E$.

PROOF. To prove the continuity at $a \in E$, let us take $\epsilon > 0$ and non-zero $b \in E$, and let us consider the mapping

$$g(x) = \beta(< x-h(a), \bar{a} >)b + h(a),$$

where β is a C^∞-function of R into R such that

$$\beta(\xi) = 0 \text{ if } |\xi| \geq \epsilon; = 1 \text{ if } \xi = 0.$$

Then, g is the composition of the following form :

$$E \xrightarrow{f_1} E \xrightarrow{f_2} R \xrightarrow{\beta} R \xrightarrow{f_3} E \xrightarrow{f_4} E,$$

where

$$f_1(x) = x - h(a), \quad f_2(x) = < x, \bar{a} >, \quad f_3(\xi) = \xi b$$

and $\qquad f_4(x) = x + h(a),$

each of which is EC^∞. Therefore, $g \in S(E)$.

We take $f \in S(E)$ such that $\phi(f) = g$. Then, $f(a) \neq a$, because if $f(a) = a$, we have

$$h(a) = hf(a) = \phi(f)h(a) = gh(a) = b + h(a),$$

which is a contradiction. Since f is continuous, there is $U \in N(E)$ such that $x - a \in U$ implies $f(x) \neq a$. Therefore, $x - a \in U$ implies

$$gh(x) = hf(x) \neq h(a),$$

which means that $\beta(< h(x)-h(a), \bar{a} >) \neq 0$. By the definition of β, we have $|< h(x) - h(a), \bar{a} >| < \varepsilon$.

As to the continuity of h itself, we need an additional assumption as follows.

(7.2.3) <u>Let</u> S <u>be a continuous</u> S-<u>category and</u> E <u>be</u> S-<u>smooth</u>. Then, the mapping h is continuous.

PROOF. To prove that h is continuous at $a \in E$, we take an arbitrary $V \in N(E)$ and we shall show that there is $U \in N(E)$ such that $x - a \in U$ implies $h(x) - h(a) \in V$. Since E is S-smooth, by (5.2.3), there exists $\beta \in S(E,R)$ such that $\beta(h(a)) = 1$ and $\{x \in E : \beta(x) \neq 0\} \subset h(a) + V$. Then, for a non-zero $b \in E$, the mapping $g \in S(E)$ defined by

$$g(x) = \beta(x)b + h(a)$$

works in the same way as in the above proof.

There is a theorem of Gelfand [1] that says, if $f : R \to E \in BS$ is weakly continuous and E is separable, then f is of the first Baire class by the strong topology and, hence, it has a point of strong continuity. Therefore, the function $h(\xi a)$ of $\xi \in R$ into E has a point of strong continuity if E is separable. Is this function continuous everywhere? We can answer this question with the following theorem.

(7.2.4) 1°. <u>Let</u> S <u>be a continuous</u> S-<u>category</u>. Then, if the mapping h is continuous at one point, it is continuous everywhere.

2°. Let S be an S-category. Assume that $S(E) \subset D_H(E)$. Then, if h is directionally, Gâteaux or Hadamard differentiable at one point, then it is differentiable everywhere in the same sense.

3°. If S is an S-category, $S(E) \subset D(E)$ and h is differentiable at one point, it is differentiable everywhere.

PROOF. 1°. Let h be continuous at a and let $b \in E$. Then, $1 + c_{b-a} \in S(E)$ and, since S is continuous, $\phi(1+c_{b-a})$ is a continuous mapping. Hence, the continuity of h at b follows from the following relation:

$$h(x) - h(b) = \phi(1+c_{b-a})h(x-b+a) - \phi(1+c_{b-a})h(a).$$

2°. Assume that $f \in D(a, \to x)$ and let $b \in E$. Then, since $\phi(1+c_{b-a}) \in D_H(E)$, we have

$$\varepsilon^{-1}[h(b+\varepsilon x) - h(b)]$$
$$= \varepsilon^{-1}[\phi(1+c_{b-a})h(a+\varepsilon x) - \phi(1+c_{b-a})h(a)]$$
$$= \phi(1+c_{b-a})'(h(a))[\varepsilon^{-1}[h(a+\varepsilon x) - h(a)]]$$
$$+ \varepsilon^{-1}r(\phi(1+c_{b-a}), h(a), h(a+\varepsilon x) - h(a))$$
$$\to \phi(1+c_{b-a})'(h(a))(h'(a,x)).$$

Hence, $h \in D(b, \to x)$ and

$$h'(b,x) = \phi(1+c_{b-a})'(h(a))(h'(a,x)).$$

If $h \in D_G(a,E)$, then, since $h'(a) \in L(E,E)$, we have $h \in D_G(b,E)$ and $h'(b) = \phi(1+c_{b-a})'(h(a)) \circ h'(a)$.

The proofs for the remaining cases are similar.

From these facts we have a result of Magill [1].

(7.2.5) Let $D(R)$ be the semigroup of all differentiable mappings of R into R. Then, every automorphism of $D(R)$ is inner.

PROOF. The semigroup $D(R)$ satisfies the assumptions of (7.2.1) and (7.2.2). Therefore, the mapping h is a homeomorphism of R onto R and, hence, it is monotone. Therefore, it is differentiable at some point. Thus, by (7.2.4), $h \in D(R)$. Hence, ϕ is inner.

The same proof as above does not apply if the dimension of E is greater than 1. In fact, in such spaces, there are homeomorphisms that are not even directionally difffferentiable at any point. The following example was given to us by I. Mycielski and S. Swierczkowski.

Example. Let $E \in$ LCS and suppose that its dimension is not less than 2. Let $\alpha : R \to R$ be a continuous and nowhere differentiable function. By the assumption, we can take $a \in E$ and $\bar{a} \in \bar{E}$ such that $a \neq 0$, $\bar{a} \neq 0$ and $<a,\bar{a}> = 0$. Now, define $f : E \to E$ by

$$f(x) = x + \alpha(<x,\bar{a}>)a.$$

It is easy to see that f is the required homeomorphism - it is not directionally differentiable at any point in the direction of x such that $<x,\bar{a}> \neq 0$.

This example suggests that, if ϕ is inner, the differentiability of h must be proved by a direct calculation. At present, we have only the following partial answers.

1°. For any $E \in$ BS, every automorphism of the semigroup $D_H(E)$ is inner. (Yamamuro [3]).

2°. Every automorphism of the semigroup $D(E)$ is inner if $E = \ell^1$ or E is an (FM)-space. (Wood-Yamamuro [1], Yamamuro [4]).

3°. Every automorphism of the semigroups $D^n(E)$ and $C^n(E)$ is inner if E is finite-dimensional. (Wood [1,2]).

For the proofs of these facts, we refer the reader to the original sources.

NOTE: In the theory of the one-parameter semigroups of non-linear mappings, there is a problem in which the differentiability is crucial, namely, the differentiability with respect to the parameter. For this problem, see Komura [1], Kato [1] and Dorroh [1]. The same problem exists in the theory of the groups of differentiable mappings, (which has some connections with the problem considered in this section - the differentiability of $h(\xi a)$ with respect to ξ is equivalent to the differentiability of the group $\{\phi(e^{\xi})\}$ with respect to the parameter). The most fundamental result on this problem is that of Bochner-Montgomery [1]. A generalization to infinite-dimensional spaces was given by Dorroh [2].

For both cases, the most comprehensive work has been done by Chernoff-Marsden [2], a part of which has appeared in [1] of the same authors.

§7.3 Near-rings of differentiable mappings

Let S be an S-category and $E \in LCS$. Then, as we have mentioned in the beginning of this chapter, $S(E)$ is a near-ring.

At first, we shall answer the question of inner automorphisms of near-rings. An automorphism of a near-ring A is a bijection $\phi : A \to A$ such that $\phi(fg) = \phi(f)\phi(g)$ and $\phi(f+g) = \phi(f) + \phi(g)$ for any $f,g \in A$.

(7.3.1) Let $E \in BS$ and A be a near-ring whose elements are continuous mappings of E into itself. If A contains $I(E)$ and $L(E) = L(E,E)$, then every automorphism of A is inner.

PROOF. By (7.2.1), ϕ has the expression (*). We have to show that $h \in A$. Let $x \in E$ be such that $h^{-1}(x) \neq 0$ and take $\bar{a} \in \bar{E}$ such that $< h^{-1}(x), \bar{a} > = 1$. Then, for arbitrary $a, b \in E$, since $a \otimes \bar{a}, b \otimes \bar{a} \in L(E) \subset A$, we have

$$\phi(a \otimes \bar{a} + b \otimes \bar{a})(x) = \phi(a \otimes \bar{a})(x) + \phi(b \otimes \bar{a})(x),$$

which, together with (*), implies

$$h(a+b) = h(a) + h(b).$$

Therefore, for any $u \in L(E)$, $\phi(u)$ is an additive mapping. Since, by the assumption, $\phi(u)$ is continuous, ϕ maps $L(E)$ into itself.

To show that ϕ maps $L(E)$ onto itself, let $u \in L(E)$. Since $\phi : A \to A$ is onto, $\phi(f) = u$ for some $f \in A$. By (*), we have $f = h^{-1}uh$, which shows that f is additive. Since f is continuous, $f \in L(E)$.

Thus, ϕ is an automorphism of the ring $L(E)$. Therefore, by a theorem of Eidelheit [1], there exists an invertible $h_1 \in L(E)$ such that $\phi(u) = h_1 u h_1^{-1}$ for all $u \in L(E)$, from which it follows that $h_1 = \alpha h$ for some $\alpha \in R$. Hence, $h \in L(E) \subset A$. Therefore, ϕ is inner.

It is an immediate consequence that every automorphism of the near-ring $S(E)$ is inner for any $E \in BS$.

Next we consider the ideals of the near-ring $S(E)$. Let A be a near-ring. A subset I of A is called an <u>ideal</u> if it is the kernel of a homeomorphism. In other words, as Blakett [1] has pointed out, I is an <u>ideal</u> if

1°. I is a normal subgroup of the additive group A;

2°. $fg \in I$ if $f \in I$ and $g \in A$;

3°. $g_1(f+g_2) - g_1 g_2 \in I$ if $f \in I$ and $g_1, g_2 \in A$.

A subset that satisfies 1° and 2° is called a <u>right ideal</u>, and a subset satisfying 1° and 3° is called a <u>left ideal</u>.

We shall consider another condition :

4°. $gf \in I$ if $f \in I$ and $g \in A$.

A subset that satisfies 1° and 4° is called a <u>left semigroup ideal</u> or <u>left (SG)-ideal</u>. If it satisfies 1°, 2° and 4°, it will be called a (SG)-ideal.

It is obvious that, if $g0 = 0$ for all $g \in A$, then 3° implies 4°. However, in fact, we can say more about this relation. As a preparation, we prove a convenient lemma.

(7.3.2) <u>Let</u> $E \in$ LCS <u>and</u> A <u>be a near-ring of mappings of</u> E <u>into</u> E. <u>If</u> $A \supset L(E)$, <u>then we have either</u> $A \supset I(E)$ <u>or</u> $f(0) = 0$ <u>for all</u> $f \in A$.

PROOF. We prove that if there is $f \in A$ such that $f(0) = a \neq 0$, then $I(E) \subset A$. If f is such, then, for any $x \in E$,

$$c_a(x) = a = f(0) = f0(x),$$

which means that $c_a = f0 \in A$. Then, for any $b \in E$ and $\bar{a} \in \bar{E}$ such that $<a, \bar{a}> = 1$,

$$c_b(x) = b = (b \otimes \bar{a})(a) = (b \otimes \bar{a}) c_a(x) \quad \text{for all } x \in E,$$

which means that $c_b \in A$. Hence, $I(E) \subset A$.

Then we have the following theorem.

(7.3.3) <u>Let</u> $E \in$ LCS <u>and</u> A <u>be a near-ring of mappings of</u> E <u>into</u> E. <u>If</u> $A \supset L(E)$, <u>every ideal is a</u> (SG)-<u>ideal</u>.

PROOF. Let I be an ideal. Since

$$g(f+g_1) - gg_1 \in I \quad \text{if } f \in I \text{ and } g_1, g \in A,$$

for $g_1 = 0$, we have

$$gf - g0 \in I \quad \text{if } f \in I \text{ and } g \in A.$$

By (7.3.2), we have two cases:

case 1: $g(0) = 0$. Then, $gf \in I$.

case 2: $I(E) \subset A$. Let $g(0) = a \neq 0$, and let $h \in I$ be any non-zero element. Then, $h(x) = b \neq 0$ for some $x \in E$. Let $\bar{a} \in \bar{E}$ be such that $<b, \bar{a}> = 1$. Then, since $a \otimes \bar{a} \in A$,

$$c_a = (a \otimes \bar{a}) c_b = (a \otimes \bar{a})(c_b + c_a) - (a \otimes \bar{a}) c_a,$$

where $c_b = hc_x \in I$. Hence, $c_a \in I$ and $c_a = g0$. Therefore, $gf \in I$.

Thus, I is a (SG)-ideal.

It is worth noting that some important subsets are not ideals but are (SG)-ideals. For example,

1°. $I(E)$: It is obvious that, if A is a near-ring and $I(E), L(E) \subset A$, then, $I(E)$ is a (SG)-ideal, and $I(E) \subset I$ for any non-trivial ideal I. However, $I(E)$ is not an ideal.

2°. $K(E)$: If A is a near-ring whose elements are bounded and continuous, then $K(E)$ is a (SG)-ideal but not an ideal.

Evidently, if $A = L(E)$, every (SG)-ideal is an ideal. However, in general, the examples of ideals are rather scarce.

3°. Let A be a near-ring whose elements are differentiable and vanish at zero. Then, $I = \{f \in A : f'(0) = 0\}$ is an ideal.

The following theorem confirms the fact that the difference between ideals and (SG)-ideals is considerable.

(7.3.4) <u>Let</u> A <u>be a near-ring of differentiable mappings of</u> $E \in LCS$. <u>If</u> $L(E) \subset A$, $f(0) = 0$ <u>for every</u> $f \in A$ <u>and every left</u> (SG)-<u>ideal is a left ideal</u>, <u>then</u> $A = L(E)$.

PROOF. Let $a \in E$ and consider the following set :

$$I_a = \{f \in A : f'(a) = 0\}.$$

Then, I_a is a left (SG)-ideal. Hence, by the assumption, it has to be left ideal. In other words,

$$g(f+g_1) - gg_1 \in I_a \text{ if } f \in I_a \text{ and } g, g_1 \in A.$$

Hence, for $g_1 = 1$, we have

$$g'(f(a)+a)(f'(a)+1) - g'(a) = 0,$$

from which it follows that

(*) $g'(f(a)+a) = g'(a)$ if $f \in I_a$ and $g \in A$.

We consider two cases :

case 1 : There is $f_0 \in I_a$ such that $f_0(a) \neq 0$.

case 2 : $f(x) = 0$ if $f \in I_x$ and $x \in E$.

In the case 1, we take $\bar{a} \in \bar{E}$ such that $< f_0(a), \bar{a} > = 1$. Then, for any $x \in E$, since $(x \bar{\otimes} \bar{a})f_0 \in I_a$, if we put $f = (x \bar{\otimes} \bar{a})f_0$ in the above (*), we have

$$g'(a+x) = g'(a),$$

which implies $g \in L(E)$ by (1.3.4). 2°.

In the case 2, since we always have $f - f'(x) \in I_x$,

$$f(x) = f'(x)(x) \quad \text{if} \quad x \in E \quad \text{and} \quad f \in A.$$

Now, let us take $\bar{a} \in \bar{E}$ and consider the functional :

$$\phi(\xi) = < f(\xi x), \bar{a} >.$$

Then, since

$$\phi'(\xi) = < f'(\xi x)(x), \bar{a} > = \frac{1}{\xi}\phi(\xi) \quad \text{if} \quad \xi \neq 0,$$

we have $\phi(\xi) = \alpha\xi$ for some constant α. Since $\bar{a}$ is arbitrary, we have $f(\xi x) = \xi f(x)$ if $\xi \in R$ and $x \in E$. Further, since f is differentiable, this implies $f(x) = f'(0)(x)$, which implies that $f \in L(E)$.

Now, we present a theory of ideals in which the differentiability plays an essential rôle. The following discussions, applied here only to the near-ring $D(E) = D(E,E)$, can also be applied to any near-ring A such that $L(E) \subset A \subset D_H(E)$.

(7.3.5) **Define** $d(f)$, $d(M)$ **and** $d^{-1}(N)$ **as follows** :

$$d(f) = \{f'(x) : x \in E\} \quad \text{for} \quad f \in D(E);$$

$$d(M) = \bigcup_{f \in M} d(f) \quad \text{for} \quad M \subset D(E);$$

$$d^{-1}(N) = \{f \in D(E) : d(f) \subset N\} \quad \text{for} \quad N \subset L(E).$$

Then,

1°. $M_1 \subset M_2$ implies $d(M_1) \subset d(M_2)$.
    $N_1 \subset N_2$ implies $d^{-1}(N_1) \subset d^{-1}(N_2)$.

2°. $d(f) = \{0\}$ if and only if $f \in I(E)$.

3°. The following three conditions are equivalent :
    (1) $f \in L(E)$; (2) $f \in d(f)$ and (3) $d(f) = \{f\}$.

4°. $M \cap L(E) \subset d(M)$ for any $M \subset D(E)$.

5°. $d(d(M)) = d(M)$ for any $M \subset D(E)$.

6°. $N \subset d^{-1}(N)$ for any $N \subset L(E)$.

7°. $d(d^{-1}(N)) = N$ for any $N \subset L(E)$.

8°. $M \subset d^{-1}(d(M))$ for any $M \subset D(E)$.

PROOF.   All proofs are routine and omitted.

We here note that, in general, equality does not necessarily hold in 8°. As the example, let E be a separable Hilbert space and put $M = (K \cap D)(E)$. Then, the mapping considered in (2.1.2). 3° belongs to $d^{-1}(M))$ but not to M.

[7.3.6] A subset $M \subset D(E)$ is said to be a d-set if $M = d^{-1}(d(M))$. An ideal is said to be a d-ideal if it is a d-set.

(7.3.7) 1°. If M is a d-set, then $d(M) = M \cap L(E)$.

2°. The following three conditions are equivalent :
    (1) M is a d-set; (2) $f \in M$ if and only if $d(f) \subset M$ and
    (3) $M = d^{-1}(N)$ for some $N \subset L(E)$.

3°. If M_1 and M_2 are d-sets and $d(M_1) = d(M_2)$, then $M_1 = M_2$.

PROOF. 1°. In view of (7.3.5) 4°, we have only to prove that $d(M) \subset M$. Since $d(M) \subset L(E)$, by (7.3.5) 6°, $d(M) \subset d^{-1}(d(M)) = M$.

2°. (1) ⇒ (2) : If $f \in M$, it follows from 1° that $d(f) \subset d(M) \subset M$. Conversely, if $d(f) \subset M$, it follows from (7.3.5) 5° that $d(f) \subset d(M)$. Hence,

$f \in d^{-1}(d(M)) = M$.

(2) ⇒ (3) : For $N = d(M)$, we have $d^{-1}(N) = d^{-1}(d(M)) \supset M$ by (7.3.5) 8°. On the other hand, if $d(f) \subset M$, since $d(g) \subset M$ whenever $g \in M$, we have

$$d(f) \subset N = d(M) = \bigcup_{g \in M} d(g) \subset M,$$

hence, $f \in M$.

(3) ⇒ (1) : It follows from (7.3.5) 7° that

$$d^{-1}(d(M)) = d^{-1}(d(d^{-1}(N))) = d^{-1}(N) = M.$$

3°. $M_1 = d^{-1}(d(M_1)) = d^{-1}(d(M_2)) = M_2$.

(7.3.8) Let $E \in BS$ and J be an ideal of the Banach algebra $L(E)$. Then, $d^{-1}(J)$ is an (SG)-ideal of $D(E)$.

PROOF. The proof is easy and omitted.

From this theorem, since $I(E) = d^{-1}(\{0\})$, $I(E)$ is a d-(SG)-ideal. On the other hand, as we have mentioned after (7.3.5), the set $K(E) \cap D(E)$ is not a d-(SG)-ideal, although it is an (SG)-ideal. In the following, we shall denote the set $K(E) \cap L(E)$ by $KL(E)$.

[7.3.9] A subset M of $D(E)$ is said to be (L)-closed if $M \cap L(E)$ is closed in $L(E)$.

The collection of all (L)-closed subsets of $D(E)$ defines a topology on $D(E)$, which is the strongest among the topologies by which the mapping $u \to u$ of $L(E)$ into $D(E)$ becomes continuous.

(7.3.10) The sets $I(E)$ and $d^{-1}(KL(E))$ are (L)-closed.

The proof of this fact is obvious. Conversely, in some cases, these are the only (L)-closed d-(SG)-ideals, as we show in the following theorem.

(7.3.11) Let $E \in BS$.

1°. Let I be an (L)-closed d-(SG)-ideal of D(E). Then, we have either $I = I(E)$ or $I \supset d^{-1}(KL(E))$.

2°. When E is a separable Hilbert space and I is an (L)-closed d-(SG)-ideal, we have either $I = I(E)$ or $I = d^{-1}(KL(E))$.

PROOF. 1°. Since $I \cap L(E)$ is a closed ideal of the Banach algebra L(E), we have either $I \cap L(E) = \{0\}$ or $I \cap L(E) \supset K(E) \cap L(E)$. From the definition of d-set and (7.3.7), 1°,

$$I = d^{-1}(d(I)) = d^{-1}(I \cap L(E)) = d^{-1}(\{0\}) = I(E)$$

or

$$I = d^{-1}(I \cap L(E)) \supset d^{-1}(KL(E)).$$

2°. When E is a separable Hilbert space, by a theorem of Calkin [1], KL(E) is the only non-zero closed ideal of L(E). Therefore, we have either $I \cap L(E) = \{0\}$ or $I \cap L(E) = KL(E)$, from which follows the required result.

NOTE: Most of the results in this section have been taken from Yamamuro [1,2]. We have an excellent bibliography on near-rings and the related topics prepared and distributed yearly by the three well-known specialists in this field : G. Betsch (Univ. of Tubingen), J.J. Malone, Jr. (Worchester Polytch.Inst.) and J.R. Clay (Univ. of Arizona).

APPENDIX 1. Sequential spaces.

A topological space E is said to be <u>sequential</u> if $a \in \bar{A} \setminus A$ for some $A \subset E$ then $x_n \to a$ for some sequence $\{x_n\} \subset A$. The importance of this notion is obvious from (1.7.1).

The proof of the following fact is taken from Averbukh-Smolyanov [2].

(1) <u>Let $E \in$ TLS be sequential. If $x_{i,j} \to x_j$ ($j=1,2,\ldots$; $i\to\infty$) and $x_j \to x_0$ ($j\to\infty$), then there are strictly increasing sequences $\{i_k\}$ and $\{j_k\}$ such that $x_{i_k, j_k} \to x_0$ ($k\to\infty$).</u>

PROOF. We may assume that $x_j = 0$ for all j. Hence, assume that $x_{i,j} \to 0$ ($j=1,2,\ldots$; $i\to\infty$). Let $a \neq 0$ and put $y_{i,j} = j^{-1}a + x_{i+j,j}$ if $j^{-1}a \neq -x_{i+j,j}$; $= j^{-1}a - x_{i+j,j}$ otherwise. Consider the set

$$A = \{y_{i,j} : i,j = 1,2,\ldots\}.$$

Then, obviously, $0 \notin A$. However, $0 \in \bar{A}$, because, for any $U \in N(E)$, if we take $V \in N(E)$ such that $V + V \subset U$, there is j such that $j^{-1}a \in V$, and then we can choose i such that $x_{i+j,j} \in V$, from which it follows that $y_{i,j} \in U$. Then, since E is sequential, $y_{i_k, j_k} \to 0$. We may take $\{i_k\}$ as strictly increasing to $+\infty$. Then, $j_k \to +\infty$. Otherwise, taking a subsequence if necessary, we can uppose $j_k = j_0 =$ constant. Then, $y_{i_k, j_k} \to j_0^{-1}a \neq 0$. It is now obvious that $x_{i_k+j_k, j_k} \to 0$.

As a consequence, we have the following fact which was used in the proof of (1.7.1)

(2) <u>Let $E \in$ TLS be sequential and $x_n \to 0$ in E. Then, there exists a subsequence $\{x_{n_k}\}$ such that $kx_{n_k} \to 0$.</u>

PROOF. Apply (1) to $\{jx_n\}$.

Let us call a subset A <u>sequentially open</u> if $a \in A$ and $x_n \to a$ imply $x_n \in A$ for large n.

(3) <u>If</u> $E \in TS$ <u>is sequential, every sequentially open set is open</u>.

PROOF. Assume that $U \subset E$ is sequentially open but not open. Then, $A = E \setminus U$ is not closed. Therefore, there is a point $a \in \bar{A} \setminus A$. Since E is sequential, $x_n \to a$ for some $x_n \in A$. Since $a \in U$, we have $x_n \in U$ for large n, which is a contradiction.

(4) <u>Every sequential</u> $E \in LCS$ <u>is bornological</u>.

PROOF. Let $A \subset E$ be an absolutely convex set which absorbs every bounded set. We shall prove that A is a neighbourhood of 0. Now, assume that, for any open set U such that $0 \in U$, we have $U \not\subset A$. Then, we can choose $x(U) \in U$ such that $x(U) \notin A$. Since $x(U) \to 0$, $x(U) \neq 0$ and E is sequential, there is a sequence $\{x(U_n)\}$ which converges to 0. Then, by (2), we have $kx(U_{n_k}) \to 0$. Since this sequence is a bounded set, $\lambda k x(U_{n_k}) \in A$ for some $\lambda > 0$ and every k. Since A is absolutely convex, $x(U_{n_k}) \in A$ for large k, which is a contradiction

Hence, if $E \in LCS$ is sequential and $F \in LCS$, any bounded linear mapping $u : E \to F$ is continuous. In the following fact the mapping does not need to be linear.

(5) <u>Let</u> $E,F \in TS$ <u>and</u> E <u>be sequential. Then, if</u> $f : E \to F$ <u>is sequentially continuous at</u> $a \in E$, <u>it is continuous there</u>.

PROOF. We show that, for any neighbourhood V of f(a) in F, $f^{-1}(V)$ is a neighbourhood of a in E. If $f^{-1}(V)$ is not a neighbourhood of a, by the same argument as in (4), we can find a sequence $\{x_n\}$ such that $x_n \to a$ and $x_n \notin f^{-1}(V)$. Since f is sequentially continuous, $f(x_n) \to f(a)$, which implies $f(x_n) \in V$ for large n, a contradiction.

Another remark on (4) is that the converse is not true. We shall show this by the method used by Averbukh-Smolyanov [2, Theorem 4.3]. We call a strict inductive limit $E = \text{ind}(E_n)$ _regular_ if $A \subset E$ is bounded in E if and only if $A \subset E_n$ for some n and bounded there.

(6) Let E be a regular inductive limit of a strictly increasing sequence $E_n \in \text{LCS}$. Then, E is not sequential.

PROOF. Take $a_1 \neq 0$ and $a_n \in E_n \setminus E_{n-1}$ for $n \geq 2$, and consider the set $A = \{a_{n,k} = n^{-1}a_1 + k^{-1}a_n : n,k = 1,2,...\}$. Then $0 \in \bar{A} \setminus A$. The proof of $0 \in \bar{A}$ is similar to the proof in (1), and $0 \notin A$ is obvious. Assume that we can choose a sequence $a_{n_i,k_i} \to 0$ and $n_i \to +\infty$. Then, $k_i^{-1}a_{n_i} \to 0$. Since $\{k_i^{-1}a_{n_i}\}$ is a bounded set, it is contained in some E_n, which is impossible.

If, for instance, $E_n \in \text{BS}$, then E is bornological. However, E is not sequential. It is well-known that E can never be metrizable. Hence, being sequential is closer to being metrizable than being bornological. Balanzat [2] gave an example of a sequential space which is not metrizable.

A Hausdorff space is said to be compactly generated or a k-space if each subset that intersects every compact set in a closed set is itself a closed set. Sequential spaces are obviously compactly generated. This notion was introduced in the book of Kelly [1] and its importance from the viewpoint of category theory has been stressed by Steenrod [1].

A calculus on compactly generated spaces has been developed by Seip [1].

APPENDIX 2. Continuity of the composition

Let $E, F, G \in$ LCS and consider the composition :

$$\text{comp} : L(E,F) \times L(F,G) \to L(E,G)$$

defined by $\text{comp}(u,v) = v \circ u$. We always assume that these spaces have the topology of uniform convergence on bounded sets. It is obvious that comp is bilinear and separately continuous.

1. <u>Continuity of the composition.</u>   In this section, we investigate the (joint) continuity, following the works by Blair [1], Maissen [1] and Keller [2].

(1.1)   <u>If</u> F <u>is normed,</u> comp <u>is continuous.</u>

PROOF.   Let $\widetilde{W} = (B,W) \in N(L(E,G))$, where

$$(B,W) = \{u \in L(E,G) : u(B) \subset W\}.$$

Put $B_1 = \{x \in F : \|x\| \leq 1\}$ and $V_1 = \{x \in F : \|x\| < 1\}$. Then, $(B, V_1) \in N(L(E,F))$, $(B_1, W) \in N(L,(F,G))$ and

$$(B_1, W) \circ (B, V_1) \subset \widetilde{W}.$$

To prove the converse, we need two lemmas.   Let $x \neq 0$ in E and consider the mapping

$$\phi_x : L(E,F) \to F \text{ defined by } \phi_x(u) = u(x).$$

(1.2) ϕ_x <u>is continuous and open.</u>

PROOF.   Let $V \in N(F)$. Then,

$$\phi_x^{-1}(V) = \{u \in L(E,F) : u(x) \in V\} = (\{x\}, V),$$

which is a neighbourhood of $L(E,F)$. Therefore, ϕ_k is continuous.

To show that it is open, let $\tilde{V} = (B,V) \in N(L(E,F))$. There is $V_1 \in N(F)$ such that $x \notin V_1$ and $V_1 \subset V$. There is $\lambda > 0$ such that $\lambda B \subset V_1$. Then, we show that $\lambda V_1 \subset \phi_x(\tilde{V})$. Now, let $y \in \lambda V_1$. We can assume that $y \neq 0$. By Bourbaki [1] II, 3.3, Prop. 4, there exists $f \in \bar{E}$ such that $f(x) = 1$ and $|f(\lambda B)| \leq 1$. Let $g : R \to F$ be an isomorphism defined by $g(\xi) = \xi y$, and put $u = g \circ f$, i.e., $u : E \xrightarrow{f} R \xrightarrow{g} F$. Then, $u \in L(E,F)$ and $u(x) = y$. Moreover, $u(\lambda B) = g(f(\lambda B)) \subset [-1,1]y \subset \lambda V_1$, i.e. $u \in (B,V_1) \subset (B,V)$. Hence, ϕ_x is open.

(1.3) <u>There exist $x \neq 0$ and $\tilde{V} \in N(L(E,F))$ such that $\phi_x(\tilde{V}) \neq F$.</u>

PROOF. Take $V \in N(F)$ such that $V \neq F$, and let $0 \neq x \in B \in B(E)$. Put $\tilde{V} = (B,V)$. Then, $\phi_x(\tilde{V}) = \{u(x) : u(B) \subset V\} \subset V$.

Now, we prove the main theorem.

(1.4) <u>If comp $: L(E,F) \times L(F,G) \to L(E,G)$ is continuous, then F is normable.</u>

PROOF. Take $x \neq 0$ in E and $\tilde{W} \in N(L(E,G))$ such that $\phi_x(\tilde{W}) \neq G$, where $\phi_x(w) = w(x)$ for all $w \in L(E,G)$. Since comp is continuous, there are $\tilde{U} \in N(L(E,F))$ and $\tilde{V} \in N(L,(F,G))$ such that $\tilde{V} \circ \tilde{U} \subset \tilde{W}$. Now, consider $\psi_x : L(E,F) \to F$ defined by $\psi_x(u) = u(x)$ for $u \in L(E,F)$. Since this is an open mapping by (1.2), the set $\psi_x(\tilde{U})$ is a neighbourhood. We prove that this is also bounded. Suppose otherwise. Then, there exists $f \in \bar{E}$ and $f(\psi_x(\tilde{U})) = R$. Take any $z \neq 0$ in G and define $g : R \to G$ by $g(\xi) = \xi z$. Then, $g \circ f \in L(F,G)$ and $\lambda(g \circ f) \in \tilde{V}$ for some $\lambda > 0$. Hence,

$$\{\xi z\} = \lambda(g \circ f)(\psi_x(\tilde{U})) = \bigcup_{u \in \tilde{U}} \lambda(g \circ f)(u(x))$$

$$= \bigcup_{u \in \tilde{U}} [\lambda(g \circ f) \circ u](x) \subset \bigcup_{w \in \tilde{W}} \{w(x)\}$$

$$= \phi_x(\tilde{W}) \neq G.$$

Since z is an arbitrary non-zero element, this is a contradiction. Therefore, F is normable.

2. **Hypocontinuity of the composition.** We first give the definition of the hypocontinuity.

[2.1] Let $E_1, E_2, F \in LCS$ and $\phi : E_1 \times E_2 \to F$ be a separately continuous bilinear mapping. Let M_i be a class of bounded sets in E_i (i=1,2). ϕ is said to be M_1-**hypocontinuous** if, for any $W \in N(F)$ and any $M_1 \in M_1$, there is $U_2 \in N(E_2)$ such that $\phi(M_1 \times U_2) \subset W$. ϕ is said to be M_2-hypocontinuous if, for any $W \in N(F)$ and any $M_2 \in M_2$, there exists $U_1 \in N(E_1)$ such that $\phi(U_1 \times M_2) \subset W$.

If ϕ is M_1- and M_2-hypocontinuous, it is called (M_1, M_2)-hypocontinuous.

We denote the class of all equicontinuous subsets of E, if it is meaningful, by $Eq(E)$.

(2.2) *The mapping* comp : $L(E,F) \times L(F,G) \to L(E,G)$ *is* $(B(L(E,F)), Eq(L(F,G)))$-*hypocontinuous.*

PROOF. We first prove the $B(L(E,F))$-hypocontinuity. Let $\tilde{W} = (B,W) \in N(L(E,G))$ and $\tilde{B} \in B(L(E,F))$. Put $B_1 = \tilde{B}(B) = \bigcup_{u \in B} u(B)$, then $B_1 \in B(F)$. Hence, for $\tilde{V} = (B_1, W)$, we have $\tilde{V} \circ B \subset W$. Next, we show the $Eq(L(F,G))$-hypocontinuity. Let $\tilde{W} = (B,W) \in N(L(E,G))$ and $\tilde{A} \in Eq(L(F,G))$. There exists $V \in N(F)$ such that $\tilde{A}(V) \subset W$. Put $\tilde{U} = (B,V)$, then we have $\tilde{A} \circ \tilde{U} \subset \tilde{W}$.

As an immediate consequence,

(2.3) *Let* $\tilde{A} \in Eq(L(F,G))$. *Then*, comp : $L(E,F) \times \tilde{A} \to L(E,G)$ *is continuous.*

3. **Sequential continuity of the composition.** A convergent sequence is not necessarily an equicontinuous set. Therefore, (2.2) and (2.3) can not be used for the discussion of the sequential continuity.

(3.1) *Let* $\{u_\sigma\} \subset L(E,F)$ *and* $\{v_\sigma\} \subset L(F,G)$ *be nets such that* $u_\sigma \to 0$ *and* $v_\sigma \to 0$. *If* $\{u_\sigma(B)\}$ *is a bounded set for each* $B \in B(E)$, *then* $v_\sigma \circ u_\sigma \to 0$.

PROOF. Let $\hat{W} = (B,W) \in N(L(E,G))$, and put $B_1 = \bigcup_\sigma u_\sigma(B)$, which, by the assumption, is a bounded set. Then, $\hat{V} = (B_1,W) \in N(L(F,G))$. Hence, there exists σ_0 such that $v_\sigma \in \hat{V}$ if $\sigma \geq \sigma_0$, which means that $(v_\sigma \circ u_\sigma)(B) \subset W$ if $\sigma \geq \sigma_0$. Therefore, $v_\sigma \circ u_\sigma \to 0$.

From this we have the following fact.

(3.2) *If* $\{u_n\} \subset L(E,F)$ *and* $\{v_n\} \subset L(F,G)$ *are sequences convergent to zero, then* $v_n \circ u_n \to 0$.

APPENDIX 3. Differentiability of inverse mappings

We shall characterize the class of spaces to which (3.2.5) can be generalized. The results that follow have been obtained by John Grunau and the author.

The main result is in the third section.

1. Spaces of type (D^{-1}).

We shall say that a space is **of type** (D^{-1}) if a generalization of (3.2.5) is possible in this space. More precisely,

[1.1] Let E,F ∈ LCS and a ∈ A ∈ $\mathcal{O}$(E). A mapping $f : A \to F$ is said to be **of type** (D^{-1}) **at** a if

(1) f is one-to-one;

(2) f is differentiable at a and f'(a) is an isomorphism,

and

(3) the inverse $f^{-1} : f(A) \to A$ is continuous at f(a).

E ∈ LCS is said to be **of type** (D^{-1}) if, for any mapping $f : A \to F$ of type (D^{-1}) at a into any F ∈ LCS, the inverse is differentiable at f(a).

Since the set f(A) may not be open, obvious changes have to be made concerning the definition of differentiability of f^{-1}. In other words, we assume in [1.2.1] that there exists a unique continuous linear mapping for which the limit there is zero when $\varepsilon \to 0$ with $a + \varepsilon x \in A$.

We know that normed spaces are of type (D^{-1}). The example of Averbukh-Smolyanov, mentioned after (3.2.5), shows that the countable products of real lines are not of type (D^{-1}). On the other hand, Suhinin's result, mentioned at the end of §3.2, says that strict inductive limits of increasing sequences of Banach spaces are of type (D^{-1}).

2. Levered spaces

The aim of this section is to prove Lemma (2.4), which will be used in the proof of (3.3) and itself is a convenient necessary condition for a space to be of type (D^{-1}).

[2.1] E ∈ LCS is said to be <u>levered</u> if, for any null sequence $\{x_n\}$ in E, there exists a sequence $\{\alpha_n\}$ such that the sequence $\{\alpha_n x_n\}$ is not convergent to zero.

Here, a <u>null</u> sequence is a sequence $\{x_n\}$ such that $\lim_{n\to\infty} x_n = 0$ and $x_n \neq 0$ for all n.

The fact that the countable product of real lines is not levered follows from the following result.

(2.2) E ∈ MLS <u>with a generating increasing sequence of semi-norms</u> $\{p_i\}$ <u>is levered if and only if some</u> p_i <u>is a norm.</u>

PROOF. Assume that p_i is a norm and $\{x_n\}$ is a null sequence. Then, we can take $\alpha_n = p_i(x_n)^{-1}$ in the above definition. Conversely, if every p_i is not a norm, we take $x_n \neq 0$ such that $p_n(x_n) = 0$. Then, for any i and for any sequence $\{\alpha_n\}$,

$$p_i(\alpha_n x_n) \leq p_n(\alpha_n x_n) = 0 \quad \text{if } n \geq i,$$

which means that $\alpha_n x_n \to 0$. Hence, E is not levered.

The following fact is used in the proof of (2.4).

(2.3) <u>If</u> E ∈ LCS <u>is not levered, there exist null sequences</u> $\{x_n\}$ <u>and</u> $\{y_n\}$ <u>such that</u>

(1) $\alpha_n y_n \to 0$ <u>for any</u> $\{\alpha_n\}$ <u>and there exist non-zero</u> $\bar{y}_n \in \bar{E}$ <u>such that</u>

$$< y_n, \bar{y}_m > = 1 \text{ if } m = n; \, = 0 \text{ if } m \neq n.$$

(2) <u>for any</u> $\{\alpha_n\}$, <u>the sequence</u> $\{\alpha_n x_n\}$ <u>is either unbounded or convergent to zero.</u>

PROOF. (1) By the assumption, there is a null sequence $\{z_n\}$ such that $\alpha_n z_n \to 0$ for any $\{\alpha_n\}$. Since such sequence can not be contained in a finite-dimensional subspace, we can choose a subsequence $\{z_{n_i}\}$ such that z_{n_i} is not in the subspace spanned by $\{z_{n_j} : 1 \leq j < i\}$. Since it is a subsequence of $\{z_n\}$,

$\alpha_i z_{n_i} \to 0$ for any $\{\alpha_i\}$. We denote this sequence by the same $\{z_n\}$.

Now, we define $\{y_n\}$ and $\{\bar{y}_n\}$ by induction. First, we put $y_1 = z_1$ and take $\bar{y}_1 \in \tilde{E}$ such that $< y_1, \bar{y}_1 > = 1$.

Assume that we have $\{y_i, \bar{y}_i : 1 \leq i \leq k\}$ which are biorthogonal. For each i, since $\alpha_n < z_n, \bar{y}_i > \to 0$ for any $\{\alpha_n\}$,

$$< z_n, \bar{y}_i > = 0 \quad \text{except for finite n's.}$$

We take the first n such that $< z_j, \bar{y}_i > = 0$ if $j \geq n$ and $1 \leq i \leq k$, and put $y_{k+1} = z_n$ for this n. Since $\{y_1, y_2, \ldots, y_{k+1}\}$ is linearly independent, there exist $\bar{z}_j \in \tilde{E}$ $(j = 1, 2, \ldots, k+1)$ such that

$$< y_i, \bar{z}_j > = 1 \text{ if } i = j; = 0 \text{ if } i \neq j.$$

We put $\bar{y}_{k+1} = \bar{z}_{k+1}$.

(2). Put

$$x_n = 2^{-2^n} y_1 + 2^{-2^{n-1}} y_2 + \ldots + 2^{-2} y_n,$$

and assume that $\{\alpha_n x_n\}$ is bounded and $\alpha_n > 0$ for all n. We prove that $\alpha_n x_n \to 0$.

Since $\{\alpha_n x_n\}$ is bounded,

$$\gamma_m = \sup_{n \geq 1} |< \alpha_n x_n, \bar{y}_m >| < + \infty,$$

and $< \alpha_n x_n, \bar{y}_m >$ is the coefficient of y_m in the expression of $\alpha_n x_n$. We denote it by $\beta_{m,n}$. Then, for $n \geq m > 1$,

$$\beta_{m,n} = \alpha_n 2^{-2^{n-m+1}}$$
$$= (\alpha_n 2^{-2^{n-m}}) 2^{-2^{n-m}}$$
$$= \beta_{m-1,n} 2^{-2^{n-m}}$$
$$\leq \gamma_{m-1} 2^{-2^{n-m}} \to 0 \quad \text{if } n \to \infty,$$

and it follows from this that $\beta_{1,n} \to 0$.

Now, take any continuous semi-norm p. Then, there exist $m_1, m_2, \ldots, m_k$ such that

$$p(y_m) = 0 \quad \text{if } n \notin \{m_1, m_2, \ldots, m_k\}.$$

Therefore,

$$p(\alpha_n x_n) \leq \sum_{i=1}^{k} p(\beta_{m_i, n} y_{m_i}) \to 0 \quad \text{if } n \to \infty.$$

Moreover,

$$p(x_n) \leq \sum_{i=1}^{k} p(2^{-2^{n-m_i+1}} y_{m_i}) \to 0.$$

Hence, $\{x_n\}$ is a null sequence and satisfies the required condition.

(2.4) **If** $E \in LCS$ **is of type** (D^{-1}), **it is levered**.

PROOF. Assume that E is not levered and take the sequences $\{x_n\}$ and $\{y_n\}$ constructed in (2.3). We define a mapping $f : E \to E$ in the same way as in the first example after (3.2.5). Namely, we put

$$x_{n,k} = (2 - \tfrac{1}{k})x_n \quad \text{and} \quad y_{n,k} = (2 - \tfrac{1}{k})y_n,$$

and define f as follows :

$$f(x_{n,k}) = x_{n,k-1} \quad (n = 1, 2, \ldots; k = 2, 3, \ldots)$$

$$f(x_n) = y_n \quad (n = 1, 2, \ldots)$$

$$f(y_{n,k}) = y_{n,k+1} \quad (n, k = 1, 2, \ldots)$$

$$f(x) = x \quad \text{if } x \notin A = \{x_{n,k}, y_{n,k} : n, k = 1, 2, \ldots\}.$$

Since f maps A onto A, it maps E onto E. Moreover, since $\{y_n\}$ is linearly independent, it is one-to-one.

To prove that f is differentiable at zero and $f'(0) = 1$, we take a null sequence $\{\varepsilon_i\}$ and a bounded sequence $\{z_i\}$. Assume that $\varepsilon_i z_i = x_{n_i, k_i}$. Then,

$\{\varepsilon_i^{-1} x_{n_i, k_i}\}$ is bounded and, hence, is a null sequence. Therefore $\{\varepsilon_i^{-1} x_{n_i, k_i - 1}\}$ is also null and

$$\varepsilon_i^{-1} f(\varepsilon_i z_i) - z_i = \varepsilon_i^{-1} x_{n_i, k_i - 1} - \varepsilon_i^{-1} x_{n_i, k_i} \to 0.$$

The other cases can be treated in the same way.

Since the inverse $g = f^{-1}$ is obviously continuous at zero, f is of type (D^{-1}) at zero. However, g is not differentiable at $f(0) = 0$. By (1.2.9), we only need to show that the identity is not the derivative of g at zero. Now, we put

$$\varepsilon_n = 2^{-2^n} \quad \text{and} \quad z_n = \varepsilon_n^{-1} y_n \quad \text{for all } n.$$

Then, $\{\varepsilon_n\}$ is null, $\{z_n\}$ is bounded and

$$\varepsilon_n^{-1} g(\varepsilon_n z_n) - z_n = \varepsilon_n^{-1} x_n - \varepsilon_n^{-1} y_n \ ,$$

where $\varepsilon_n^{-1} y_n \to 0$. However,

$$< \varepsilon_n^{-1} x_n, \bar{y}_1 > = 1 \quad \text{for all } n,$$

which implies that $\varepsilon_n^{-1} x_n \not\to 0$.

Hence, E is not of type (D^{-1}).

3. Boundedly levered spaces

[3.1] $E \in LCS$ is said to be __boundedly levered__ if, for each null sequence $\{x_n\}$ in E, there exists a sequence $\{\alpha_n\}$ such that the sequence $\{\alpha_n x_n\}$ is bounded but not convergent to zero.

Obviously, boundedly levered spaces are levered and also braked in the sense of Wilansky[1]. Normed spaces are boundedly levered because we can take $\alpha_n = \|x_n\|^{-1}$ in the above definition. It is also easy to see that the strict inductive limit of an increasing sequence of Banach spaces is boundedly levered.

(3.2) **If** $E \in$ LCS **is boundedly levered, it is of type** (D^{-1}).

PROOF. Let $F \in$ LCS, $a \in A \in \mathcal{O}(E)$ and $f : A \to F$ be of type (D^{-1}) at a. Considering the mapping

$$x \mapsto f'(a)^{-1}[f(a+x) - f(a)],$$

we can assume that $E = F$, $a = 0$ and $f'(0) = 1$. Assume that the identity is not the derivative of $g = f^{-1}$ at zero. Then, there exist a null sequence $\{\varepsilon_n\}$, a bounded sequence $\{y_n\}$ and a continuous semi-norm p such that

$$p(\varepsilon_n^{-1} g(\varepsilon_n y_n) - y_n) > \delta$$

for some positive number δ. Then, since

$$\varepsilon_n^{-1} g(\varepsilon_n y_n) - y_n = x_n - \varepsilon_n^{-1} f(\varepsilon_n x_n)$$

for

$$x_n = \varepsilon_n^{-1} g(\varepsilon_n y_n),$$

the sequence $\{x_n\}$ does not contain a bounded subsequence. Since g is continuous at zero and $\varepsilon_n y_n \to 0$, we have $\varepsilon_n x_n \to 0$. By the assumption, we can find $\{\alpha_n\}$ such that $\{\alpha_n \varepsilon_n x_n\}$ is bounded but not convergent to zero.

We prove that $\alpha_n \varepsilon_n \to 0$. If this does not converge to zero, there exists a subsequence $\{\alpha_{n_i} \varepsilon_{n_i}\}$ such that $\inf_{i \geq 1} |\alpha_{n_i} \varepsilon_{n_i}| > 0$. Then, since $\{\alpha_{n_i} \varepsilon_{n_i} x_{n_i}\}$ is bounded, $\{x_{n_i}\}$ must be bounded, a contradiction.

Now, we always have the following relation:

$$\alpha_n \varepsilon_n x_n = \alpha_n \varepsilon_n y_n - \alpha_n r(f, 0, \alpha_n^{-1}(\alpha_n \varepsilon_n x_n)),$$

and the right-hand side converges to zero but the left-hand side does not.

Next, we shall prove that the converse is also true.

(3.3) **If** $E \in$ LCS **is of type** (D^{-1}), **it is boundedly levered**.

PROOF. Assume that E is of type (D^{-1}) and not boundedly levered. Then, there exists a null sequence $\{e_n\}$ such that, for any $\{\alpha_n\}$, the sequence $\{\alpha_n e_n\}$ is either unbounded or convergent to zero. By (2.4), we can find $\{\rho_n\}$ such that $\rho_n \to +\infty$ and $\{\rho_n e_n\}$ does not converge to zero.

Now, we define a mapping $f : E \to E$ in the same way as in the second example after (3.2.5). First, we choose sequences $\{\lambda_n\}$ and $\{\mu_n\}$ such that

$$\lambda_n \to 0, \quad \mu_n \to 0 \quad \text{and} \quad \lambda_n \mu_n^{-1} - 1 = \rho_n,$$

and put

$$\lambda_{n,k} = (2 - \tfrac{1}{k})\lambda_n \quad \text{and} \quad \mu_{n,k} = (2 - \tfrac{1}{k})^{-1}\mu_n \quad (n,k = 1,2,\ldots).$$

Now, we define $f : E \to E$ as follows :

$$f(\lambda_{n,k} e_n) = \lambda_{n,k-1} e_n \qquad (n = 1,2,\ldots; \; k = 2,3,\ldots)$$

$$f(\lambda_n e_n) = \mu_n e_n \qquad (n = 1,2,\ldots)$$

$$f(\mu_{n,k} e_n) = \mu_{n,k+1} e_n \qquad (n,k = 1,2,\ldots)$$

$$f(x) = x \quad \text{if} \quad x \notin A = \{\lambda_{n,k} e_n, \mu_{n,k} e_n\}.$$

Then, it is easy to see that f is of type (D^{-1}) at zero and $f'(0) = 1$. However, $g = f^{-1}$ is not differentiable at $f(0) = 0$, because

$$\mu_n^{-1} g(\mu_n e_n) - e_n = \mu_n^{-1} \lambda_n e_n - e_n = \rho_n e_n.$$

Hence, E is not of type (D^{-1}), a contradiction.

Thus, what remains is to find all locally convex Hausdorff spaces which are boundedly levered. For example,

(3.4) <u>Let</u> E ∈ LCS <u>and</u> E_w <u>be the space</u> E <u>equipped with the weak topology.</u> <u>Then, if</u> E_w <u>is boundedly levered, every weakly convergent sequence is convergent</u> <u>to the same limit by the original topology.</u> <u>If</u> E ∈ NLS, <u>the converse is also true.</u>

PROOF. Assume that E_w is boundedly levered and $\{x_n\}$ is a null sequence in E_w. If $x_n \not\to 0$ in E, there exist a continuous semi-norm p, a subsequence $\{x_{n_i}\}$ and $\delta > 0$ such that

$$p(x_{n_i}) > \delta \qquad \text{for all } i.$$

Since $x_{n_i} \to 0$ in E_w, we can find $\{\alpha_i\}$ such that $\alpha_i > 0$ and $\{\alpha_i x_{n_i}\}$ is bounded but not null. However, this is impossible, because

$$p(\alpha_i x_{n_i}) > \alpha_i \delta \qquad \text{for all } i$$

and $\{\alpha_i\}$ is not bounded.

The converse for $E \in$ NLS is obvious, because E is boundedly levered.

SYMBOLS

a	:	an element.
A	:	a subset, usually an open set.
b	:	an element.
B	:	a subset, usually a bounded set.
$B_\alpha(a)$	:	the open ball $\{x : \|a-x\| < \alpha\}$.
B_α	:	$B_\alpha(0)$.
$\mathcal{B}(E)$	:	the class of all bounded subsets of E.
BS	:	the class of all real Banach spaces.
c	:	an element.
c_a	:	the constant mapping : $c_a(x) = a$ for all x.
(c_0)	:	the Banach space of all real sequences $\{\xi_n\}$ such that $\xi_n \to 0$ with the norm $\|\{\xi_n\}\| = \sup\limits_{n \geq 1} |\xi_n|$.
$C^n(a,Y)$	:	the set of all mappings of a neighbourhood of a into Y that are n-times continuously differentiable at a.
$C^n(X,Y)$	:	the set of all n-times continuously differentiable mappings of X into Y.
Co(M)	:	the convex cover of M.
$|\text{Co}|(M)$	:	the absolutely convex cover of M.
$D(a,Y;\to x)$	:	the set of all mappings of a neighbourhood of a into Y that are differentiable at a in the direction of x.
$D(a,Y;\to X)$	:	$\bigcap\limits_{x \in X} D(a,Y;\to x)$.
$D(A,Y;\to X)$	:	the set of all mappings of A into Y that are directionally differentiable at every point of A in the direction of any point in X.
$D_G(a,Y)$	:	the set of all mappings of a neighbourhood of a into Y that are Gâteaux differentiable at a.

$D_G(X,Y)$	:	the set of all Gâteaux differentiable mappings of X into Y.
$D_H(a,Y)$	:	the same as $D_G(a,Y)$ where Gâteaux is replaced by Hadamard.
$D_H(X,Y)$	:	the same as $D_G(X,Y)$ where Gâteaux is replaced by Hadamard.
$D^n(a,Y)$	:	the set of all mappings of a neighbourhood of a into Y that are n-times Fréchet differentiable at a.
$D^n(X,Y)$	:	the set of all n-times Fréchet differentiable mappings of X into Y.
$D(X,Y)$	:	$D^1(X,Y)$.
$D_M(X,Y)$	:	the same as above with M-differentiability.
E	:	a real topological linear space.
$\bar{E}$	:	the conjugate space of E with the topology of uniform convergence on each bounded set.
f	:	a mapping.
F	:	a real topological linear space.
g	:	a mapping.
G	:	a real topological linear space.
h	:	a mapping.
H	:	a subset.
i	:	a positive integer.
I	:	an interval in the real line, or, an ideal.
inv.	:	the mapping $u \mapsto u^{-1}$.
Isom(E,F)	:	the set of all topological linear isomorphisms of E onto F.
j	:	a positive integer.
J	:	an ideal.
k	:	a positive integer, or, a mapping related to compactness.
$K(E)$	:	the class of all compact subsets of E.
$K_s(E)$	:	the class of all sequentially compact subsets of E.

$K_p(E)$	:	the class of all precompact subsets of E.
$K(E,F)$	:	the class of all B-compact mappings of E into F.
$K_p(E,F)$	:	the class of all B-precompact mappings (maps every bounded set into a precompact) of E into F.
$L(E,F)$	:	the set of all continuous linear mappings of E into F with the topology of uniform convergence on each bounded set.
$L(E)$	:	$L(E,E)$.
LCS	:	the class of all locally convex Hausdorff spaces.
m	:	a positive integer.
M	:	a subset.
MLS	:	the class of all locally convex metric linear spaces.
$M(E)$	:	a class of subsets of E such that all singletons belong to it.
$M[0,1]$	:	the Banach space of all essentially bounded measurable real functions on $[0,1]$.
n	:	a positive integer.
N	:	a subset.
NLS	:	the class of all real normed linear spaces.
$N(E)$	:	the class of all circled open neighbourhood of 0 of E. When E is locally convex, the elements are assumed to be absolutely convex.
$O(E)$	:	the class of all open subsets of E.
O	:	the class of all open subsets of all topological linear spaces.
p	:	a semi-norm.
p_B	:	a semi-norm on $L(E,F)$ defined by $p_B(u) = \sup_{x \in B} p(u(x))$.
$P(E)$	:	the set of all continuous semi-norms of $E \in$ LCS.
q	:	a semi-norm, or a mapping.
Q	:	a set.

r	:	a mapping.
r(f,a,x)	:	the remainder of f at a.
R	:	the set of all real numbers with the usual topology.
R^+	:	the set of all positive real numbers.
R^n	:	n-dimensional Euclidean space.
s	:	a mapping.
S(U,V)	:	the set of morphisms of U into V in an S-category S.
t	:	a mapping.
TLS	:	the class of all topological linear Hausdorff spaces.
u	:	an element of L(E,F).
U	:	an open subset, mostly, an element of $N(E)$.
v	:	an element of L(E,F).
V	:	an open subset, mostly, an element of $N(F)$.
w	:	an element of L(E,F).
W	:	an open set.
x	:	an element.
X	:	a subset.
y	:	an element.
Y	:	a subset.
z	:	an element.
Z	:	a subset.
$\bar{M}$	:	the closure of M.
[a,b]	:	$\{(1-\xi)a+\xi b : 0 \leq \xi \leq 1\}$.
(a,b)	:	$\{(1-\xi)a+\xi b : 0 < \xi < 1\}$.
$<x,\bar{x}>$	:	the value of $\bar{x} \in \bar{E}$ at $x \in E$, where $E \in$ LCS.

REFERENCES

This only covers the works directly connected with the topics discussed in this lecture. Nashed [1] has excellent references from the functional analytic viewpoint, and Eells [1,2] also has excellent references from the geometric viewpoint.

R. Abraham and J. Robbin

1. Transversal mappings and flows, W.A. Benjamin, 1967.

A. Alexiewicz

1. On differentiation of vector-valued functions, Studia Math., 11 (1949), 187-196.

A. Alexiewicz and W. Orlicz

1. On the differentials in Banach spaces, Ann. Soc. Polon. Math., 25 (1952), 95-99.

E. Asplund and R.T. Rockafellar

1. Gradients of convex functions, Trans. Amer. M.S., 139 (1969), 443-467. MR 39#1968.

V.I. Averbukh and O.G. Smolyanov

1. The theory of differentiation in linear topological spaces, Russian M. Survey, 22:6 (1967), 201-258.

2. The various definitions of the derivatives in linear topological spaces, ibid, 23:4 (1968), 67-113. MR 39#7424.

M. Balanzat

1. La différential en les espacios métricas affines, Matem. Notae, 9(1949), 29-51.

2. La differentielle d'Hadamard-Fréchet dans les espaces vectorielles topologiques, C.R. Acad. Sci. Paris, 251 (1960), 2459-2461 MR 23#A3478.

R.G. Bartle

1. The Elements of Real Analysis, John Wiley and Sons, 1967.

A. Bastiani

1. Applications différentiables et variétes différentiables de dimension infinie, J. Analy. M., 13 (1965), 1-114.

C. Bessega

1. Every infinite dimensional Hilbert space is diffeomorphic with its unit sphere, Bull. l'Acad. Polonaise Sci., 14 (1966), 27-31.

E. Binz

1. Ein Differenzierbarkeitbegriff in limitierten Vektorräumen, Comm. M. Helv., 41 (1966), 137-156.

E. Binz and W. Meier-Solfrian

1. Zur Differenzialrechnung in limitierten Vektorräumen, ibid., 42 (1967), 285-296.

D.W. Blackett

1. Simple and semi-simple near-rings, Proc. Amer. M.S., 4 (1953), 772-785.
2. Simple near-rings of differentiable transformations, Proc. Amer. M.S., 7 (1956), 599-606.

A. Blair

1. Continuity of multiplication in operator algebra, Proc. Amer. M.S., 6 (1955), 209-220.

S. Bochner and D. Montgomery

1. Groups of differentiable and real or complex analytic transformations, Ann. Math., 46 (1945), 685-694.

T.S. Bolis

1. Smooth partitions of unity in some infinite dimensional manifolds (to appear).

R. Bonic

1. Four brief examples concerning polynomials on certain Banach spaces, J.Diff. Geometry, 2 (1968), 391-392. MR 39#6075.

R. Bonic and J. Frampton

1. Differentiable functions on certain Banach spaces, Bull. Amer. M.S., 71 (1965), 393-395.

2. Smooth functions on Banach manifolds, J. Math. Mec., 15 (1966), 877-898.

R. Bonic, J. Frampton and A. Tromba

1. Λ-manifolds, J. Func. Analysis, 3 (1969), 310-320. MR 38#6635.

R. Bonic and F. Reis

1. A characterization of Hilbert spaces, Anais Acad. Brasileira ciências, 38 (1966), 239-241.

N. Bourbaki

1. Espaces vectoriels topologiques, Hermann, 1953-1955.

W. Bucher

1. See A. Fröhlicher

2. Différentiabilité de la composition et complétitude de certains espaces functionnels, Comment. Math. Helv., 43 (1968), 256-288.

P. Chernoff and J. Marsden

1. On continuity and smoothness of group actions, Bull. Amer. M.S., 76 (1970), 1044-1049.

2. Hamiltonian Systems and Quantum Mechanics (to appear).

J.A. Clarkson

1. Uniformly convex spaces, Trans. Amer. M.S., 40 (1936), 396-414.

J.F. Colombeau

1. Calcul différentiel dans les espaces bornologiques, Thése doctorat, Bordeaux, 1970.

S.H. Cox, Jr. and S.B. Nadler, Jr.

1. Supremum norm differentiability, Ann. Soc. Math. Polonaise, 15 (1971), 127-131.

D.F. Cudia

1. Rotundity, Proc. Symp. Pure Math., vol. 7 (1963), 73-97.
2. The geometry of Banach spaces, Smoothness, Trans. Amer. M.S., 110 (1964), 284-314.

J. Daneš and J. Kolomý

1. On the continuity and differentiability properties of convex functionals, comment. Math. Univ. Carolinae, 9 (1968), 329-350.

P.J. Daniell

1. The derivative of a functional, Bull. Amer. M.S., 25 (1919), 414-416.

M.M. Day

1. Normed Linear Spaces, Springer, 1957.
2. Strict convexity and smoothness of normed spaces, Trans. Amer. M.S., 78 (1955), 516-528.

S.F.L. de Foglio

1. La différentielle au sens d'Hadamard dans les espaces L vectorielles, Portugal M., 19 (1960), 165-184.

J.D. de Pree and J.A. Higgins

1. Collectively compact sets of linear operators, Math. Z., 115 (1970), 366-370. MR 41#9019.

J. Dieudonné

1. Foundations of Modern Analysis, Academic Press, 1960.
2. Treatise on Analysis, Academic Press, 1970.

M.D. Donsker and J.L. Lions

1. Fréchet-Voltera variational equations, boundary value problems and function space integrals, Acta M., 108 (1962), 147-228.

J.R. Dorroh

1. Semigroups of non-linear transformations, Michigan M.J., 12 (1965), 317-320.
2. Local groups of differentiable transformations, Math. Ann., 192 (1971), 243-249.

E. Dubinsky

1. Differential equations and differential calculus in Montel spaces, Trans. Amer. M.S., 110 (1964), 1-21.

J. Eells, Jr.,

1. A setting for global analysis, Bull. Amer. M.S., 72 (1966), 751-807.
2. Fredholm structures, Proc. Symp. Pure Math., vol. 18 (1970), 62-85.

V.G. Fajans

1. Isomorphisms of semigroups of affine transformations, Siberian M.J., 11 (1970), 154-158.

P.L. Falb and M.Q. Jacobs

1. On differentials in locally convex spaces, J. Diff. Eq., 4 (1968), 444-459.

Ky Fan

1. Sur quelques notions fundamentals de l'analyse générale, J.M. Pures Appl., 21 (1942), 289-368.

H. Federer

1. Geometric Measure Theory, Grundlehren der math. Wiss., Band 153, Springer, 1969.

H.R. Fisher

1. Differentialkalkül für nicht metrische Strukturen, Ann. Acad. Sci. Fenn., 247 (1957), MR 19#869.

2. Limesräume, Math. Ann., 137 (1959), 269-303.

K. Floret and J. Wloka

1. Einführung in die Theorie der lokalkonvexen Räume, Lecture Notes in Math., vol. 56, Springer, 1968.

J. Frampton

See R. Bonic.

M. Fréchet

1. La notion de différentielle dans l'analyse générale, C.R. Acad. Sci. Paris, 180 (1925), 806-809.

2. Sur la notion de différentielle dans l'analyse générale, J.M. Pures Appl. 16 (1937), 233-250.

A. Fröhlicher and W. Bucher

1. Calculus in Vector Spaces without Norms, Lecture Notes in Math., vol. 30, Springer, 1966.

R. Gâteaux

1. Sur les fonctionelles continue et les fonctionelles analytiques, C.R. Acad. Sci. Paris, 157 (1913), 325-327.

I. Gelfand

1. Abstrakte Funktionen und linearen Operatoren, Recueil Math., 4 (1938) (46), 235-284.

J.R. Giles

1. On a characterization of differentiability of the norm of a normed linear space, J. Austral. M.S., 12 (1971), 106-114.
2. On a differentiability condition for reflexivity of a Banach space, J. Austral. M.S., 12 (1971), 393-396.
3. A Banach space with support homeomorphism is reflexive, Bull. Austral. M.S., 5 (1971), 187-189.
4. A non-reflexive Banach space has non-smooth third conjugate space.

V. Goodman

1. Quasi-differentiable functions on Banach spaces, Proc. Amer. M.S., 30 (1971), 367-370.

L.M. Graves

1. Topics in the functional calculus, Bull. Amer. M.S., 41 (1935), 641-662.

A. Grothendieck

1. Sur certains espaces de functions holomorphes, J. Reine Angew. M., 192 (1953), 35-64 and 77-95.

J. Hadamard

1. Sur les transformations ponctuelles, Bull. Soc. Math. France, 34 (1906), 71-84.
2. La notion de différentielle dans l'enseignement, scripta Univ. Ab. Hierosolymitanarum, Jerusalem, 1 (1923), 3.

J.A. Higgins

See J.D. De Pree.

W. Hurewicz and H. Wallman

1. Dimension Theory, Princeton Math. Series, vol.4, 1962.

D.H. Hyers

1. Linear topological spaces, Bull. Amer. M.S., 51 (1945), 1-21.

K. Iseki

1. Implicit functions on locally convex topological linear spaces, Proc. Japan. Acad., 41 (1965), 147-149.

M.Q. Jacobs

See P.L. Falb.

M.I. Kadec

1. Conditions for the differentiability of a norm in a Banach space, Uspehi M. Nauk, 20 (1965), 183-187.

T. Kato

1. Note on the differentiability of non-linear semigroups, Proc. Symp. Pure Math., vol. 16 (1970), 91-94.

H.H. Keller

1. Differenzierbarkeit in topologischen Vektorräumen, Comment. Math. Helv., 38 (1964), 308-320. MR 29#3858
2. Räume stetiger multilinearer Abbildungen als Limesräume, Math. Ann., 159 (1965), 259-270.
3. Über Probleme, die bei einer Differentialrechnung in topologishen Vektorräumen auftreten, Festband sum 70 Geburtstag von Rolf Nevalinna, Springer, 1966, 49-57.

J.L. Kelley

1. General Topology, D. van Nostrand, New York, 1955.

A. Khintichine

1. Recherches sur la structure des fonctions measurables, Fund.Math., 9 (1926), 212-279.

J. Kijowski and W. Szczyrba

1. On differentiability in an important class of locally convex spaces, Studia M., 30 (1968), 247-257. MR 38#1524.

V. Klee

1. Mappings into normed linear spaces, Fund. M., 49 (1960), 25-63.

J. Kolomý

1. On the differentiability of mappings in functional spaces, Comment. Math. Univ. Carolinae, 8 (1967), 315-329.
2. See J. Daneš.

J. Kolomý and V. Zizler

1. Remarks on the differentiability of mappings in normed linear spaces, Comment. Math. Univ. Carolinae, 8 (1967), 691-704. MR 38#2596.

Y. Kōmura

1. Differentiability of non-linear semigroups, J. Math. Soc. Japan, 21 (1969), 375-402.

G. Köthe

1. Topological Vector-Spaces, I, Springer-Verlag, New York Inc., 1969.

M.A. Krasnoselskii

1. Topological Methods in the Theory of Nonlinear Operator Equations, Pergamon, New York, 1964.

J. Kurzweil

1. On approximation in real Banach spaces, Studia M., 14 (1953), 214-231.

S. Lang

1. Introduction to Differentiable Manifolds, John Wiley and Sons, 1967.

E.B. Leach and J.H.M. Whitfield

1. Differentiable functions and rough norms on Banach spaces, Proc. Amer. M.S., 33 (1972), 120-126.

M. Leduc

1. Jauges différentiables et partitions de l'unite, Séminaire Choquet, Initiation á l'analyse, no.12 (1964/65), 223-232.

T.E. Leonard and K. Sundaresan

1. Geometry of Lebesgue-Bochner function spaces - Smoothness, (to appear).

P. Lévy

1. Sur les fonctions de lignes implicites, Bull. Soc. Math. France, 48 (1920), 13-27.

2. Leçons d'Analyse Fonctionelle, Gauthiers-Villars, 1922.

J.L. Lions

See M.D. Donsker.

J.W. Lloyd

1. Differentiable mappings on topological vector spaces, Studia M., (to appear).

2. Inductive and projective limits of smooth topological vector spaces, Bull. Austral. M.S., 6 (1972), 227-240.

3. Smooth partitions of unity on manifolds, (to appear).

K.D. Magill, Jr.

1. Automorphisms of the semigroup of all differentiable functions, Glasgow, M.J., 8 (1967), 63-66.

B. Maissen

1. Über Topologien im Endomorphismenraum eines topologischen Vektorraumes, Math. Ann., 151 (1963), 283-285.

J. Marcinkiewicz and A. Zygmund

1. On the differentiability of functions and the summability of trigonometrical series, Fund. M., 26 (1936), 1-43.

G. Marinescu

1. Asupra differentialei si derivatei in sptiile normate, Acad. R.P. Romine, Bull. Sti. Sect. Sti. Mat. Fiz., 6 (1954), 213-219.
2. Différentelles de Gâuteaux et Fréchet dans les espaces localement convexes, Bull. Math. Soc. R.P. Roumaine (N.S.) I (49) (1957), 77-86. MR 20#1188.

J. Marsden

See P. Chernoff.

S. Mazur

1. Über konvexe Mengen in linearen normierten Räumen, Studia M., 4 (1933), 70-84.

D. Meeus

1. Sur la derivée d'une fonction entre parties d'espaces localement convexes, C.R. Acad. Sci. Paris, 271 (1970), 1250-1253.

W. Meier-Solfrian

See E. Binz.

A.D. Michal

1. Differential calculus in linear topological spaces, Proc. Nat. Acad. Sci. U.S.A., 24 (1938), 340-342.
2. Differentials of functions with argument and values in topological abelian groups, ibid., 26 (1940), 356-359.

J.B. Miller

1. Generalized Gâteaux and Fréchet derivatives in convolution algebras, Proc. Cambridge Philos. Soc., 59 (1963), 707-718.

D. Montgomery

See S. Bochner.

D. Montgomery and L. Zippin

1. Topological Transformation Groups, Interscience Pub., 1955.

R.H. Moore

1. Differentiability and convergence for compact non-linear operators, J.M. Analy., 16 (1966), 65-72.

S. Nadler, Jr.

1. The idempotents of a semigroup, Amer. M.Monthly, 70 (1963), 996-997.
2. Differentiable retractions in Banach spaces, Tohoku M.J., 19 (1967), 400-405.
3. See S.H. Cox, Jr.

M.Z. Nashed

1. Differentiability and related properties of nonlinear operators: Some aspects of the role of differentials in nonlinear functional analysis, Nonlinear Functional Analysis and Applications (Proc. Advanced Sem., Math. Res. Center, Univ. of Wisconsin, Madison, Wis., 1970), 103-309. Academic Press, New York, 1971.

W. Orlicz

1. See A. Alexiewicz.

K.J. Palmer

1. On the complete continuity of differentiable mappings, J. Austral. M.S., 9 (1969), 441-444. MR#4715

A. Pelcyński

1. On almost diffeomorphic Banach spaces, Indag.M., 30 (1968), 202-208.

Jean-Paul Penot

1. Calcul différentiel dans les espaces vectoriels topologiques, Studia Math. 47 (1973) (to appear).

B.J. Pettis

1. Differentiation in Banach spaces, Duke M.J., 5 (1939), 254-269.

R.R. Phelps

1. Subreflexive normed linear spaces, Arch.M., 8 (1957), 444-450.

M.M. Rao

1. Smoothness of Orlicz spaces, Indag. M., (1965), 671-690.

2. Characterizing Hilbert space by smoothness, ibid., (1967), 132-135.

3. Notes on characterizing Hilbert space by smoothness and smooth Orlicz spaces, J.M. Anal. Appl. 37 (1972), 228-234.

F. Reis

See R. Bonic.

G. Restrepo

1. Differentiable norms in Banach spaces, Bull. Amer. M.S., 70 (1964), 413-414.

W.C. Rheinboldt

1. Local mapping relations and global implicit function theorems, Trans. Amer. M.S., 138 (1969), 183-198. MR 39#1990.

C.E. Rickart

1. One-to-one mappings of rings and lattices, Bull. Amer. M.S., 54 (1948), 758-764.

J. Robbin

See A. Abraham.

R.T. Rockafellar

    See E. Asplund.

S. Saks

1. Theory of the Integral, Warszawa-Lwow, 1937.

D.S. Sal'ko

1. On the theory of implicit functions in locally convex spaces, Uspehi Mat. Nauk. 24 (1969) (2), 235-236. MR 39#4672.

H.H. Schaefer

1. Topological Vector Spaces, Springer, 1971.

F.W. Schäfke

1. Differenzierbare Abbildungen (mimeographed), Univ. Köln, 1966.

J.T. Schwartz

1. Non-linear Functional Analysis, Gordon and Breach, 1969.

L. Schwartz

1. Un lemma sur la dérivation des fonctions vectorielles d'une variable réelle, Ann. Inst. Fourier, 2 (1950), 17-18.
2. Cours d'Analyse Mathematique, Hermann, 1967.

J. Sebastião e Silva

1. Le calcul différentiel et integral dans les espaces localement convexes, réels ou complexes, I, II., Atti. Acad. Lincei Rend., 20:6 (1956), 743-750, 21:1-2 (1956), 40-46. MR 19#561.
2. Concietos de função differenciável em espacos localmente conexos, Publ. Centro Estudes Matem. Lisboa, Inst. Alta Cultura, (1957). MR 21#283.
3. Les espaces á bornes et la notion de fonction différentiable, Colloque sur l'analyse fonctionelle, Louvain, 1961, 57-61. MR 25#3354.

U. Seip
1. Kompakt erzeugte Vektorräume und Analysis, Lecture Note in Math., no.273, 1972. Springer-Verlag. Berlin, Heidelberg, New York.

B. Sjöberg
1. Derivatives in topological vector spaces, Nordisk Mat. Tidskr., 14 (1966), 87-96.

O.G. Smolyanov
See V.I. Averbukh.

V.L. Smulyan
1. Sur la derivabilité de la norme dans l'espaces de Banach, Doklady Akad. Nauk SSSR, 27 (1940), 643-648.

M. Sova
1. General theory of differentiation in linear topological spaces, Czech. Math. J., 14 (1964), 485-508.
2. Conditions of differentiability in linear topological spaces, ibid., 16 (1966), 339-362.

I.N. Spatz
1. Smooth Banach algebras, Proc. Amer. M.S., 22 (1969), 328-329. MR 39#6079.

N.E. Steenrod
1. A convenient category of topological spaces, Michigan Math. J., 14 (1967), 133-152.

W. Stepanoff
1. Über totale differenzierbarkeit, Math. Ann., 90 (1923), 318-320.
2. Sur les conditions de l'existence de la différentielle totale, Rec. Math. Soc. Math. Moscow, 32 (1925), 511-526.

M.F. Suhinin

1. Two versions of a theorem on the differentiation of the inverse function in certain linear topological space (Russian), Vestnik Moskow Univ. Ser. 1, 24 (1969), 34-38.

K. Sundaresan

1. Smooth Banach spaces, Math. Ann., 173 (1967), 191-199.
2. Spaces of continuous functions into a Banach space, Studia Math. (to appear).
3. Some geometric properties of the unit cell in spaces $C(X,B)$, Bull. Acad. Polon. Sci., 19 (1971), 1007-1012.
4. See I.E. Leonard.

W. Szczyrba

1. See J. Kijowski.
2. Differentiation in locally convex spaces, Studia M., 39 (1971), 289-306.

A. Tromba

See R. Bonic and J. Frampton.

S.L. Trojanski

1. Smoothly reflexive Banach spaces, Studia M., 37 (1971), 173-180.

M.M. Vainberg

1. Various topics of the differential calculus in linear spaces, Uspehi Mat. Nauk, 4 (1952), 55-102.
2. Variational Methods for the Study of Non-linear Operators, Holden-Day, 1964.

V. Volterra

1. Sopra le funzioni che dipendeno da altre funzioni, Atti Reale Accad. Lincei Rend., (4) 3:2 (1887), 97-105, 141-146, 153-158.

H. Wallman

See W. Hurewicz.

J.C. Wells

1. Differentiable functions on c_0, Bull. Amer. M.S., 75 (1969), 117-118. MR 38#2590.

2. C^1-partitions of unity on nonseparable Hilbert space, Bull. Amer. M.S., 77 (1971), 804-807. MR 43 #5560.

I.H.M. Whitfield

1. Differentiable functions with bounded nonempty support on Banach spaces, Bull. Amer. M.S., 72 (1966), 145-146.

2. See E.B. Leach.

J. Wloka

See K. Floret.

G.R. Wood

1. On the semigroup of D^k mappings on (FM)-spaces, (to appear).

2. On the semigroup of C^k self maps of R^n, (to appear).

G.R. Wood and S. Yamamuro

1. On the semigroup of differentiable mappings, (II), Glasgow M.J., 13 (1972), 122-128.

S. Yamamuro

1. Ideals and homomorphisms in some near-rings, Proc. Japan Acad., 42 (1966), 427-432.

2. A note on d-ideals in some near-algebras, J. Austral. M.J., 7 (1967), 129-134.

3. On the semigroup of Hadamard differentiable mappings, J. Austral. M.S., 14 (1972), 329-334.

4. On the semigroup of differentiable mappings on Montel spaces, Tôhoku M.J., 24 (1972), 359-370.

K. Yoshida

1. Functional Analysis, Springer, 1965.

W.H. Young

1. The fundamental theorems of differential calculus, Cambridge Tracts in Math. and Math. Phys., no.11, 1909.

L. Zippin

See D. Montgomery.

V. Zizler

See J. Kolomý.

A. Zygmund

1. Smooth functions, Duke M.J., 12 (1945), 47-87.

INDEX

	(Section)
absolutely continuous	(6.2)
almost C^n-diffeomorphic	(5.2)
automorphism of near-rings	(7.3)
automorphism of semigroups	(7.2)
ẞ-compact (-precompact) mapping	(2.1)
bounded variation	(6.2)
boundedly levered	Appendix 3
chain rule	(1.2)
compact mapping	(2.1)
compactly generated	Appendix 1
completely bounded operator	(3.3)
composition mapping	(3.1)
composition property	(1.2)
constant mapping	(7.2)
continuous S-category	(5.1)
C_p-mapping	(3.4)
d-ideal	(7.3)
differentiability of Lipschitz mappings	(6.1)
differentiability of L^p-norms	(4.5)
differentiability of supremum norms	(4.4)
directional derivative	(1.1)
equicontinuously differentiable	(1.9)
Fréchet derivative	(1.2)
Fréchet differentiability of semi-norms	(4.2)
Gâteaux derivative	(1.2)

Hadamard derivative	(1.2)
Hadamard differentiability of semi-norms	(4.1)
Hadamard-Levy theorem	(3.5)
higher derivatives	(1.8)
higher derivatives of semi-norms	(4.3)
higher order chain rules	(1.8)
ideals of near-rings	(7.3)
idempotents	(7.1)
inductive limit	(1.6)
inverse mapping theorem	(3.4)
inverse operation	(3.1)
inverse to Taylor's theorem	(1.8)
Khintichine's theorem	(6.2)
levered	Appendix 3
lipschitzian	(1.4)
local injection theorem	(3.5)
local surjection theorem	(3.5)
(LS)-space	(1.7)
Magill's theorem	(7.2)
Mazur's theorem	(4.1)
M-derivative	(1.2)
mean value theorems	(1.3)
(M_1, M_2)-preserving	(3.2)
open mapping theorem	(3.5)
partial derivative	(1.11)
p-bounded mapping	(3.3)
peak point	(4.4)

peaking function	(4.4)
precompact mapping	(2.1)
projective topology	(1.5)
pseudodifferentiable	(6.1)
quasi-differentiable	(1.2)
Restrepo's theorem	(4.2)
S-approximation property	(5.3)
S-category	(5.1)
separably valued	(6.1)
sequential	(1.7)
$S(E,R)$-topology	(5.2)
S-normal	(5.3)
S-partition of unity	(5.3)
split C^n-imbedding theorem	(3.5)
split C^n-projection theorem	(3.5)
S-smooth mapping	(5.1)
S-smooth space	(5.2)
Stepanoff mapping	(6.2)
strongly continuous	(2.1)
strongly differentiable	(1.7)
strongly S-smooth space	(5.2)
Taylor's theorem	(1.8)
uniformly differentiable	(1.10)
weakly dense	(6.1)
Whitfield's theorem	(5.2)
Zygmund's theorem	(6.2)

Vol. 215: P. Antonelli, D. Burghelea and P. J. Kahn, The Concordance-Homotopy Groups of Geometric Automorphism Groups. X, 140 pages. 1971. DM 16,-

Vol. 216: H. Maaß, Siegel's Modular Forms and Dirichlet Series. VII, 328 pages. 1971. DM 20,-

Vol. 217: T. J. Jech, Lectures in Set Theory with Particular Emphasis on the Method of Forcing. V, 137 pages. 1971. DM 16,-

Vol. 218: C. P. Schnorr, Zufälligkeit und Wahrscheinlichkeit. IV, 212 Seiten. 1971. DM 20,-

Vol. 219: N. L. Alling and N. Greenleaf, Foundations of the Theory of Klein Surfaces. IX, 117 pages. 1971. DM 16,-

Vol. 220: W. A. Coppel, Disconjugacy. V, 148 pages. 1971. DM 16,-

Vol. 221: P. Gabriel und F. Ulmer, Lokal präsentierbare Kategorien. V, 200 Seiten. 1971. DM 18,-

Vol. 222: C. Meghea, Compactification des Espaces Harmoniques. III, 108 pages. 1971. DM 16,-

Vol. 223: U. Felgner, Models of ZF-Set Theory. VI, 173 pages. 1971. DM 16,-

Vol. 224: Revêtements Etales et Groupe Fondamental. (SGA 1). Dirigé par A. Grothendieck XXII, 447 pages. 1971. DM 30,-

Vol. 225: Théorie des Intersections et Théorème de Riemann-Roch. (SGA 6). Dirigé par P. Berthelot, A. Grothendieck et L. Illusie. XII, 700 pages. 1971. DM 40,-

Vol. 226: Seminar on Potential Theory, II. Edited by H. Bauer. IV, 170 pages. 1971. DM 18,-

Vol. 227: H. L. Montgomery, Topics in Multiplicative Number Theory. IX, 178 pages. 1971. DM 18,-

Vol. 228: Conference on Applications of Numerical Analysis. Edited by J. Ll. Morris. X, 358 pages. 1971. DM 26,-

Vol. 229: J. Väisälä, Lectures on n-Dimensional Quasiconformal Mappings. XIV, 144 pages. 1971. DM 16,-

Vol. 230: L. Waelbroeck, Topological Vector Spaces and Algebras. VII, 158 pages. 1971. DM 16,-

Vol. 231: H. Reiter, L^1-Algebras and Segal Algebras. XI, 113 pages. 1971. DM 16,-

Vol. 232: T. H. Ganelius, Tauberian Remainder Theorems. VI, 75 pages. 1971. DM 16,-

Vol. 233: C. P. Tsokos and W. J. Padgett. Random Integral Equations with Applications to stochastic Systems. VII, 174 pages. 1971. DM 18,-

Vol. 234: A. Andreotti and W. Stoll. Analytic and Algebraic Dependence of Meromorphic Functions. III, 390 pages. 1971. DM 26,-

Vol. 235: Global Differentiable Dynamics. Edited by O. Hájek, A. J. Lohwater, and R. McCann. X, 140 pages. 1971. DM 16,-

Vol. 236: M. Barr, P. A. Grillet, and D. H. van Osdol. Exact Categories and Categories of Sheaves. VII, 239 pages. 1971. DM 20,-

Vol. 237: B. Stenström, Rings and Modules of Quotients. VII, 136 pages. 1971. DM 16,-

Vol. 238: Der kanonische Modul eines Cohen-Macaulay-Rings. Herausgegeben von Jürgen Herzog und Ernst Kunz. VI, 103 Seiten. 1971. DM 16,-

Vol. 239: L. Illusie, Complexe Cotangent et Déformations I. XV, 355 pages. 1971. DM 26,-

Vol. 240: A. Kerber, Representations of Permutation Groups I. VII, 192 pages. 1971. DM 18,-

Vol. 241: S. Kaneyuki, Homogeneous Bounded Domains and Siegel Domains. V, 89 pages. 1971. DM 16,-

Vol. 242: R. R. Coifman et G. Weiss, Analyse Harmonique Non-Commutative sur Certains Espaces. V, 160 pages. 1971. DM 16,-

Vol. 243: Japan-United States Seminar on Ordinary Differential and Functional Equations. Edited by M. Urabe. VIII, 332 pages. 1971. DM 26,-

Vol. 244: Séminaire Bourbaki - vol. 1970/71. Exposés 382-399. IV, 356 pages. 1971. DM 26,-

Vol. 245: D. E. Cohen, Groups of Cohomological Dimension One. V, 99 pages. 1972. DM 16,-

Vol. 246: Lectures on Rings and Modules. Tulane University Ring and Operator Theory Year, 1970-1971. Volume I. X, 661 pages. 1972. DM 40,-

Vol. 247: Lectures on Operator Algebras. Tulane University Ring and Operator Theory Year, 1970-1971. Volume II. XI, 786 pages. 1972. DM 40,-

Vol. 248: Lectures on the Applications of Sheaves to Ring Theory. Tulane University Ring and Operator Theory Year, 1970-1971. Volume III. VIII, 315 pages. 1971. DM 26,-

Vol. 249: Symposium on Algebraic Topology. Edited by P. J. Hilton. VII, 111 pages. 1971. DM 16,-

Vol. 250: B. Jónsson, Topics in Universal Algebra. VI, 220 pages. 1972. DM 20,-

Vol. 251: The Theory of Arithmetic Functions. Edited by A. A. Gioia and D. L. Goldsmith VI, 287 pages. 1972. DM 24,-

Vol. 252: D. A. Stone, Stratified Polyhedra. IX, 193 pages. 1972. DM 18,-

Vol. 253: V. Komkov, Optimal Control Theory for the Damping of Vibrations of Simple Elastic Systems. V, 240 pages. 1972. DM 20,-

Vol. 254: C. U. Jensen, Les Foncteurs Dérivés de $\varprojlim$ et leurs Applications en Théorie des Modules. V, 103 pages. 1972. DM 16,-

Vol. 255: Conference in Mathematical Logic – London '70. Edited by W. Hodges. VIII, 351 pages. 1972. DM 16,-

Vol. 256: C. A. Berenstein and M. A. Dostal, Analytically Uniform Spaces and their Applications to Convolution Equations. VII, 130 pages. 1972. DM 16,-

Vol. 257: R. B. Holmes, A Course on Optimization and Best Approximation. VIII, 233 pages. 1972. DM 20,-

Vol. 258: Séminaire de Probabilités VI. Edited by P. A. Meyer. VI, 253 pages. 1972. DM 22,-

Vol. 259: N. Moulis, Structures de Fredholm sur les Variétés Hilbertiennes. V, 123 pages. 1972. DM 16,-

Vol. 260: R. Godement and H. Jacquet, Zeta Functions of Simple Algebras. IX, 188 pages. 1972. DM 18,-

Vol. 261: A. Guichardet, Symmetric Hilbert Spaces and Related Topics. V, 197 pages. 1972. DM 18,-

Vol. 262: H. G. Zimmer, Computational Problems, Methods, and Results in Algebraic Number Theory. V, 103 pages. 1972. DM 16,-

Vol. 263: T. Parthasarathy, Selection Theorems and their Applications. VII, 101 pages. 1972. DM 16,-

Vol. 264: W. Messing, The Crystals Associated to Barsotti-Tate Groups: With Applications to Abelian Schemes. III, 190 pages. 1972. DM 18,-

Vol. 265: N. Saavedra Rivano, Catégories Tannakiennes. II, 418 pages. 1972. DM 26,-

Vol. 266: Conference on Harmonic Analysis. Edited by D. Gulick and R. L. Lipsman. VI, 323 pages. 1972. DM 24,-

Vol. 267: Numerische Lösung nichtlinearer partieller Differential- und Integro-Differentialgleichungen. Herausgegeben von R. Ansorge und W. Törnig, VI, 339 Seiten. 1972. DM 26,-

Vol. 268: C. G. Simader, On Dirichlet's Boundary Value Problem. IV, 238 pages. 1972. DM 20,-

Vol. 269: Théorie des Topos et Cohomologie Etale des Schémas. (SGA 4). Dirigé par M. Artin, A. Grothendieck et J. L. Verdier. XIX, 525 pages. 1972. DM 50,-

Vol. 270: Théorie des Topos et Cohomologie Etale des Schémas. Tome 2. (SGA 4). Dirigé par M. Artin, A. Grothendieck et J. L. Verdier. V, 418 pages. 1972. DM 50,-

Vol. 271: J. P. May, The Geometry of Iterated Loop Spaces. IX, 175 pages. 1972. DM 18,-

Vol. 272: K. R. Parthasarathy and K. Schmidt, Positive Definite Kernels, Continuous Tensor Products, and Central Limit Theorems of Probability Theory. VI, 107 pages. 1972. DM 16,-

Vol. 273: U. Seip, Kompakt erzeugte Vektorräume und Analysis. IX, 119 Seiten. 1972. DM 16,-

Vol. 274: Toposes, Algebraic Geometry and Logic. Edited by. F. W. Lawvere. VI, 189 pages. 1972. DM 18,-

Vol. 275: Séminaire Pierre Lelong (Analyse) Année 1970-1971. VI, 181 pages. 1972. DM 18,-

Vol. 276: A. Borel, Représentations de Groupes Localement Compacts. V, 98 pages. 1972. DM 16,-

Vol. 277: Séminaire Banach. Edité par C. Houzel. VII, 229 pages. 1972. DM 20,-

Vol. 278: H. Jacquet, Automorphic Forms on GL(2). Part II. XIII, 142 pages. 1972. DM 16,-

Vol. 279: R. Bott, S. Gitler and I. M. James, Lectures on Algebraic and Differential Topology. V, 174 pages. 1972. DM 18,-

Vol. 280: Conference on the Theory of Ordinary and Partial Differential Equations. Edited by W. N. Everitt and B. D. Sleeman. XV, 367 pages. 1972. DM 26,-

Vol. 281: Coherence in Categories. Edited by S. Mac Lane. VII, 235 pages. 1972. DM 20,-

Vol. 282: W. Klingenberg und P. Flaschel, Riemannsche Hilbertmannigfaltigkeiten. Periodische Geodätische. VII, 211 Seiten. 1972. DM 20,-

Vol. 283: L. Illusie, Complexe Cotangent et Déformations II. VII, 304 pages. 1972. DM 24,-

Vol. 284: P. A. Meyer, Martingales and Stochastic Integrals I. VI, 89 pages. 1972. DM 16,-

Vol. 285: P. de la Harpe, Classical Banach-Lie Algebras and Banach-Lie Groups of Operators in Hilbert Space. III, 160 pages. 1972. DM 16,-

Vol. 286: S. Murakami, On Automorphisms of Siegel Domains. V, 95 pages. 1972. DM 16,-

Vol. 287: Hyperfunctions and Pseudo-Differential Equations. Edited by H. Komatsu. VII, 529 pages. 1973. DM 36,-

Vol. 288: Groupes de Monodromie en Géométrie Algébrique. (SGA 7 I). Dirigé par A. Grothendieck. IX, 523 pages. 1972. DM 50,-

Vol. 289: B. Fuglede, Finely Harmonic Functions. III, 188. 1972. DM 18,-

Vol. 290: D. B. Zagier, Equivariant Pontrjagin Classes and Applications to Orbit Spaces. IX, 130 pages. 1972. DM 16,-

Vol. 291: P. Orlik, Seifert Manifolds. VIII, 155 pages. 1972. DM 16,-

Vol. 292: W. D. Wallis, A. P. Street and J. S. Wallis, Combinatorics: Room Squares, Sum-Free Sets, Hadamard Matrices. V, 508 pages. 1972. DM 50,-

Vol. 293: R. A. DeVore, The Approximation of Continuous Functions by Positive Linear Operators. VIII, 289 pages. 1972. DM 24,-

Vol. 294: Stability of Stochastic Dynamical Systems. Edited by R. F. Curtain. IX, 332 pages. 1972. DM 26,-

Vol. 295: C. Dellacherie, Ensembles Analytiques, Capacités, Mesures de Hausdorff. XII, 123 pages. 1972. DM 16,-

Vol. 296: Probability and Information Theory II. Edited by M. Behara, K. Krickeberg and J. Wolfowitz. V, 223 pages. 1973. DM 20,-

Vol. 297: J. Garnett, Analytic Capacity and Measure. IV, 138 pages. 1972. DM 16,-

Vol. 298: Proceedings of the Second Conference on Compact Transformation Groups. Part 1. XIII, 453 pages. 1972. DM 32,-

Vol. 299: Proceedings of the Second Conference on Compact Transformation Groups. Part 2. XIV, 327 pages. 1972. DM 26,-

Vol. 300: P. Eymard, Moyennes Invariantes et Représentations Unitaires. II. 113 pages. 1972. DM 16,-

Vol. 301: F. Pittnauer, Vorlesungen über asymptotische Reihen. VI, 186 Seiten. 1972. DM 18,-

Vol. 302: M. Demazure, Lectures on p-Divisible Groups. V, 98 pages. 1972. DM 16,-

Vol. 303: Graph Theory and Applications. Edited by Y. Alavi, D. R. Lick and A. T. White. IX, 329 pages. 1972. DM 26,-

Vol. 304: A. K. Bousfield and D. M. Kan, Homotopy Limits, Completions and Localizations. V, 348 pages. 1972. DM 26,-

Vol. 305: Théorie des Topos et Cohomologie Etale des Schémas. Tome 3. (SGA 4). Dirigé par M. Artin, A. Grothendieck et J. L. Verdier. VI, 640 pages. 1973. DM 50,-

Vol. 306: H. Luckhardt, Extensional Gödel Functional Interpretation. VI, 161 pages. 1973. DM 18,-

Vol. 307: J. L. Bretagnolle, S. D. Chatterji et P.-A. Meyer, Ecole d'été de Probabilités: Processus Stochastiques. VI, 198 pages. 1973. DM 20,-

Vol. 308: D. Knutson, λ-Rings and the Representation Theory of the Symmetric Group. IV, 203 pages. 1973. DM 20,-

Vol. 309: D. H. Sattinger, Topics in Stability and Bifurcation Theory. VI, 190 pages. 1973. DM 18,-

Vol. 310: B. Iversen, Generic Local Structure of the Morphisms in Commutative Algebra. IV, 108 pages. 1973. DM 16,-

Vol. 311: Conference on Commutative Algebra. Edited by J. W. Brewer and E. A. Rutter. VII, 251 pages. 1973. DM 22,-

Vol. 312: Symposium on Ordinary Differential Equations. Edited by W. A. Harris, Jr. and Y. Sibuya. VIII, 204 pages. 1973. DM 22,-

Vol. 313: K. Jörgens and J. Weidmann, Spectral Properties of Hamiltonian Operators. III, 140 pages. 1973. DM 16,-

Vol. 314: M. Deuring, Lectures on the Theory of Algebraic Functions of One Variable. VI, 151 pages. 1973. DM 16,-

Vol. 315: K. Bichteler, Integration Theory (with Special Attention to Vector Measures). VI, 357 pages. 1973. DM 26,-

Vol. 316: Symposium on Non-Well-Posed Problems and Logarithmic Convexity. Edited by R. J. Knops. V, 176 pages. 1973. DM 18,-

Vol. 317: Séminaire Bourbaki - vol. 1971/72. Exposés 400-417. IV, 361 pages. 1973. DM 26,-

Vol. 318: Recent Advances in Topological Dynamics. Edited by A. Beck, VIII, 285 pages. 1973. DM 24,-

Vol. 319: Conference on Group Theory. Edited by R. W. Gatterdam and K. W. Weston. V, 188 pages. 1973. DM 18,-

Vol. 320: Modular Functions of One Variable I. Edited by W. Kuyk. V, 195 pages. 1973. DM 18,-

Vol. 321: Séminaire de Probabilités VII. Edité par P. A. Meyer. VI, 322 pages. 1973. DM 26,-

Vol. 322: Nonlinear Problems in the Physical Sciences and Biology. Edited by I. Stakgold, D. D. Joseph and D. H. Sattinger. VIII, 357 pages. 1973. DM 26,-

Vol. 323: J. L. Lions, Perturbations Singulières dans les Problèmes aux Limites et en Contrôle Optimal. XII, 645 pages. 1973. DM 42,-

Vol. 324: K. Kreith, Oscillation Theory. VI, 109 pages. 1973. DM 16,-

Vol. 325: Ch.-Ch. Chou, La Transformation de Fourier Complexe et L'Equation de Convolution. IX, 137 pages. 1973. DM 16,-

Vol. 326: A. Robert, Elliptic Curves. VIII, 264 pages. 1973. DM 22,-

Vol. 327: E. Matlis, 1-Dimensional Cohen-Macaulay Rings. XII, 157 pages. 1973. DM 18,-

Vol. 328: J. R. Büchi and D. Siefkes, The Monadic Second Order Theory of All Countable Ordinals. VI, 217 pages. 1973. DM 20,-

Vol. 329: W. Trebels, Multipliers for (C, α)-Bounded Fourier Expansions in Banach Spaces and Approximation Theory. VII, 103 pages. 1973. DM 16,-

Vol. 330: Proceedings of the Second Japan-USSR Symposium on Probability Theory. Edited by G. Maruyama and Yu. V. Prokhorov. VI, 550 pages. 1973. DM 36,-

Vol. 331: Summer School on Topological Vector Spaces. Edited by L. Waelbroeck. VI, 226 pages. 1973. DM 20,-

Vol. 332: Séminaire Pierre Lelong (Analyse) Année 1971-1972. V, 131 pages. 1973. DM 16,-

Vol. 333: Numerische, insbesondere approximationstheoretische Behandlung von Funktionalgleichungen. Herausgegeben von R. Ansorge und W. Törnig. IV, 296 Seiten. 1973. DM 24,-

Vol. 334: F. Schweiger, The Metrical Theory of Jacobi-Perron Algorithm. V, 111 pages. 1973. DM 16,-

Vol. 335: H. Huck, R. Roitzsch, U. Simon, W. Vortisch, R. Walden, B. Wegner und W. Wendland, Beweismethoden der Differentialgeometrie im Großen. IX, 159 Seiten. 1973. DM 18,-

Vol. 336: L'Analyse Harmonique dans le Domaine Complexe. Edité par E. J. Akutowicz. VIII, 169 pages. 1973. DM 18,-

Vol. 337: Cambridge Summer School in Mathematical Logic. Edited by A. R. D. Mathias and H. Rogers. IX, 660 pages. 1973. DM 42,-

Vol. 338: J. Lindenstrauss and L. Tzafriri, Classical Banach Spaces. IX, 243 pages. 1973. DM 22,-

Vol. 339: G. Kempf, F. Knudsen, D. Mumford and B. Saint-Donat, Toroidal Embeddings I. VIII, 209 pages. 1973. DM 20,-

Vol. 340: Groupes de Monodromie en Géométrie Algébrique. (SGA 7 II). Par P. Deligne et N. Katz. X, 438 pages. 1973. DM 40,-

Vol. 341: Algebraic K-Theory I, Higher K-Theories. Edited by H. Bass. XV, 335 pages. 1973. DM 26,-

Vol. 342: Algebraic K-Theory II, "Classical" Algebraic K-Theory, and Connections with Arithmetic. Edited by H. Bass. XV, 527 pages. 1973. DM 36,-

If you have any concerns about our products,
you can contact us on
ProductSafety@springernature.com

In case Publisher is established outside the EU,
the EU authorized representative is:
**Springer Nature Customer Service Center GmbH
Europaplatz 3, 69115 Heidelberg, Germany**

Printed by Libri Plureos GmbH
in Hamburg, Germany